Herbert William Conn

The Living World

Whence it Came and Whither it is Drifting

Herbert William Conn

The Living World
Whence it Came and Whither it is Drifting

ISBN/EAN: 9783337094829

Printed in Europe, USA, Canada, Australia, Japan

Cover: Foto ©ninafisch / pixelio.de

More available books at **www.hansebooks.com**

WHENCE IT CAME AND WHITHER
IT IS DRIFTING

A REVIEW OF THE SPECULATIONS CONCERNING THE ORIGIN AND
SIGNIFICANCE OF LIFE AND OF THE FACTS KNOWN IN RE-
GARD TO ITS DEVELOPMENT, WITH SUGGESTIONS
AS TO THE DIRECTION IN WHICH THE
DEVELOPMENT IS NOW TENDING

BY

H. W. CONN

PROFESSOR OF BIOLOGY IN WESLEYAN UNIVERSITY
AUTHOR OF "EVOLUTION OF TO-DAY"

G. P. PUTNAM'S SONS

NEW YORK   LONDON
27 WEST TWENTY-THIRD ST.   27 KING WILLIAM ST., STRAND
The Knickerbocker Press
1891

PREFACE.

THE world of to-day is the product of the past and the foundation of the future. We who live to-day can only judge of the past by inference from the present, and our only knowledge of the future is obtained by observing the direction in which the world seems to be trending as judged from its history. Since life is the crowning phenomenon of nature its study leads us closest to the hidden secrets of creation. The history of man has long been studied; the history of nature leading to man has only just begun to claim attention. It is not long that man has seen that history could be learned from other sources than written records, and it is only a few decades since he has conceived the idea of reading the history of life from nature itself. For half a century now have scientists been trying to pierce the mysteries connected with the life of the past. The history has not yet by any means been read, but enough has been revealed to make a sketch of it, as it appears to the biologist to-day, not a profitless undertaking.

Of the history which can be written to-day much is uncertain, much is no more than pure hypothesis and speculation variously attested by miscellaneous facts; but, on the other hand, much is more firmly established than any recorded history handed down to us in manuscript. Some parts of the history of life must rank among our most certain facts of knowledge. In the following outline of this history the endeavor will be made to separate carefully those parts which are speculative from those which are based on a more firm foundation, and to give to each part the value it deserves.

It is well to note at the outset that the facts as collected prove to us that the past history has been a continuous one, and one in which the facts follow each other with such logical consecutiveness that it may be regarded as a logical whole. The history of life has not been one of isolated facts, but a continuous flow, in which each step is foreshadowed by the one immediately preceding, and in its turn foretells the next one. This condition of things has led to a *development* in the life of the world and to a series of facts which science has termed evolution. Whatever be our philosophical understanding of this term, the facts, so far as life is concerned, are beyond question. The history of life has been one of the development of forms and types from each other, or rather from common centres. This law we call organic evolution. The facts upon which organic evolution is based are beyond controversy, although there may be still some dispute as to the interpretation of the facts. No one whose judgment means anything in the scientific world questions that the history of life has been one of growth and development of types from common centres. To this extent the theory of evolution may be considered as proved beyond dispute. In the following pages, therefore, the development of types from centres will be taken for granted. This is the most easily described by the terms development and evolution, and the history of life will therefore be outlined as one of *evolution*, without implying by the use of this word anything as to the manner of this evolution or as to its philosophical significance.

H. W. CONN.

MIDDLETOWN, July, 1891.

CONTENTS.

THE LIVING WORLD.

CHAPTER I.

INTRODUCTION—SOURCES OF BIOLOGICAL HISTORY.

To discover the history of life and to predict its future is perhaps the chief problem of biological science. This is the basis of the discussion of organic evolution ; this gives meaning to the schemes of classification of animals and plants ; this is the inspiration of investigation in embryology and geology, of experiments on spontaneous generation, and is indeed the object of biological discussion generally. Biologists are in every way trying to discover how life arose and how it developed into its present forms. Every source of evidence that can bear on the question is probed—the microscope, the chemist's retort, and the geologist's hammer, each lending its assistance. There is of course no written history on the subject, no recorded sources from which to draw anything except a few of the most recent facts. The evidence from which the history is to be drawn must be taken wholly from such accidental records as nature presents charily to our inspection. The indefiniteness and complexity of this sort of evidence make the problem a very difficult one, and it is not to be wondered at if as yet the complete history cannot be told. But if the record is disjointed, there is, on the other hand, one advantage which history drawn from such a source presents. Recorded history is frequently intentionally falsified, and more often written so as to give personal impressions of the historian by incorrectly stating the facts, and thus making it impossible to determine the truths of which detailed record is written. There is no such falsifying possible to the record given of the history of life written in nature. We must believe nature is true, or give up all hope of knowledge. While, then, the complexity of the record makes the interpretation difficult, the impossibility of nature making a false record gives to the conclusions that are reached a certain security, of which the biologist is proud and which is the justification of his claim that he deals only with facts.

There are several distinct sources in the realm of scientific facts from which the student endeavors to read the history of life. Recorded history of any sort, of course, tells us almost nothing. A few facts concerning the stability of recent species we do succeed in learning from the monumental records of Egypt, and a few facts in the history of man are written, but this is all.

Evidence from Fossils.

The most valuable source of evidence from which we can trace the history of life of past ages is that of fossil remains. From the beginning of the deposition of the stratified rocks it has been constantly happening that animals and plants which have died have been preserved in muds and sediments. These sediments subsequently harden into rocks, and the buried animals become fossils. From the fossils thus buried we can get many glimpses of the life of early times, and we have only to examine the fossils preserved in the successive layers of rocks to be able to formulate more or less of a detailed history of the life of geological ages.

Now this source of history has some decided advantages. In the first place, when dealing with fossils we are dealing with actual animals, and not with a record simply. When we find a fossil we know something of the size, shape, and appearance of the actual animal or plant that once lived, and thus we need not confine ourselves to general ideas of type. We implicitly trust the evidence given us by fossils, and do not have to ask if some modifying circumstances have deceived us. When we find a fossil oyster it is impossible to question that oysters were in existence at the time of the deposition of the rocks in which the fossil was found, so that the history drawn from this source is positive, so far as it goes.

But, on the other hand, although fossils give us abundant details, the history which can be drawn from them is sometimes nothing more than a history of details, lacking in the perspective view that makes up a true history. In the first place, taking all of the layers of stratified rocks together, we do not by any means have a continuous record of geological ages. The layers of rocks are the positive points of history, but they are separated by blank periods of which no history has been preserved, and of whose duration even we can get no estimate, except that they were extremely long. These periods must forever be blank, and it was during these periods unfortunately that the greatest changes in the history of life occurred. Secondly, fossils can give us history of those animals alone which have had a hard skeleton. The hard parts of animals may readily be preserved as fossils, but not the soft parts. Consequently, of the orders of animals having no skeletons, fossils can tell us almost nothing. Unfortunately, too, all of the early forms of life agree in having slight hard parts or none. The Protozoa, Cœlenterata, and Vermes have left fossil records in only a few cases. We are now learning further that the animals which formed the beginning of the large groups were usually without skeletons. The early cœlenterates, mollusks, and even vertebrates probably had no skeletons hard enough for preservation. Again, it is evident that the early representatives of any type were few in number, and their chance of preservation was, therefore, slight. Hence it follows that fossils can only in exceptional cases tell us anything of the early history and development of groups. Further, since the skeletons alone are preserved, fossils can tell us little of the development of internal structure of animals, and this is, after all, probably, the principal feature of importance, for in most cases it seems to have been changes in the anatomy of soft parts which have inaugurated all of the larger departures in animal history. Again, the rocks in which fossils have been deposited have been frequently subjected to the modifying effects of heat and pressure (metamorphosis). Sometimes this metamorphosis has completely obliterated whatever fossils may have been in them originally, and very commonly it has so obscured their structure as to render it impossible to determine very definitely what they originally were. It is readily seen that, the older the rocks the more liable they will be to such metamorphoses, and hence the farther back into history we go, the less definite becomes the record. Of the most recent epochs, quite a complete history is obtainable, but as we go back the record becomes less and less sure, and finally it stops altogether. Of the earliest history of life fossils give us absolutely no trace, and it is even true that fossils give us no satisfactory record of the early history of any of the groups of animals.

From all this it will appear that although fossil history is very definite where it is obtainable, it will at best be a disjointed history abounding in details at some points and lacking at others. Some small steps in the history will be given in the most minute particulars, because of the abundance of animals with skeletons to represent them, while other immense epochs will be entirely unrecorded from the lack of proper conditions to produce and preserve fossils representing the period. Moreover, from the great metamorphosis of the rocks of the older periods, the study of fossils will give us absolutely no record

of the early history of life and nothing of its origin.
As we shall soon see, the earliest fossil history represents a period when there was already a well-devel-
oped fauna, and this could of course not have been
the beginning of life.

Evidence from Embryology and Anatomy.

Some other source of evidence, then, is needed to
assist the record of fossils, and especially to aid in
giving the history of the earliest periods. At this
point we find a second important source of his-
torical evidence in the study of embryology. This
we may compare to a written history, since the history of the race is written in the development of the
individual, and since, like written history, it is open
to certain forms of false statement. Embryology
gives us a *record* of past animals, and does not, like
paleontology, offer the animals themselves for inspection.

The foundation of the value of embryology as a
source of history is based upon the fact that the
embryology of an animal repeats in outline the history of its ancestors. This fact is sometimes called
the first biological law. It was announced early in
the century by Von Bear and Agassiz, and has subsequently been investigated and confirmed by scores
of naturalists until it is safe to say that there is not
a fact in biological science that rests upon surer
basis. It is hardly necessary here to enter into a
discussion of the facts upon which this law is based.
For our purpose, it is sufficient to state that the law

is universally accepted in the biological world, and we may assume with confidence that it will be a faithful guide in the study of the history of life, at least in so far as animals are concerned.

Embryology, as a source of history, offers certain advantages over the record of fossiliferous rocks, and at the same time it is open to many difficulties of its own. If the embryology of an animal repeats the past history of its ancestors, and if all animals have descended from various common points of origin in the past, it would seem that we should only have to study the embryology of a few animals from each great type of the life in order to determine accurately the history of life in the past ages. There are two facts, however, which render such a simple proceeding impossible.

First: The embryological history of an animal lasts a few days or a few weeks; the past history of life has taken thousands of centuries. It is plain that the embryological history of animals is too short, and the past history of the life of the world is too long to make it possible that the former should retain a record of all the details of the latter. In the epitome of the history which is presented to us by the development of an animal, we find only a few of the salient points in the history preserved, which we may suppose represent such epochs of the past as have been of the greatest importance, and have therefore left the most lasting impression. In the embryological history we find that details are left entirely unrecorded. The study of an embryonic stage can give no idea of the actual appearance of the ancestors which it represents. Size, habits, shape, are entirely unknown, and frequently we cannot tell whether the early animals possessed a skeleton or a protective armor, and numerous other points are left entirely without even a suggestion. Embryology does give us general points of the fundamental structure of the early types, does give us clearly an outline of the changes through which animals passed in their early history, does give us an epitome of the early development and growth of living things. Such a history is much like a history of mankind which should give an outline of the development of social relations and political institutions, which should tell that the family relation was superseded by that of the tribe, this by the absolute monarchy and the constitutional monarchy, and finally by the government by the people. Such a history, dwelling more or less at length upon these various forms of government and showing their relations as well as their modes of origin from each other, would be much like the history which the study of embryology gives us of the early life. A history giving no names, no ideas of nations, we should regard as a poor history of nations and people, but it might be a fair history of mankind in general. So embryology deals in types and their relations and origins, but tells very little of actual animals. In dealing with a past ancestral stage thus represented, we do not have the same sort of satisfaction as in handling a fossil. We do not know just what sort of an animal was really represented by the stage in question, but we do have the satisfaction of feeling that we are dealing with a type which was of importance in the history of animals, and therefore of more far-reaching significance than any fossil or score of fossils that we could pick out of the rocks. The relations expressed by corresponding stages of different animals are very suggestive, and a short study of embryology will tell more of the true history of animals than the collection of thousands of fossils.

Second: The second fact that renders the study of history from this source a difficult matter is the very common falsification of the true record. It is undoubtedly the law that animals tend to repeat past history in their embryology, but it is also true that not infrequently various modifying circumstances occur to prevent the history being correctly recorded. The true record of past history is thus commonly obscured by numerous departures caused by the conditions surrounding the embryo, and this of course renders the reading of the history a difficult matter. The embryologist has, however, methods of correcting these errors in large measure, and of reaching trustworthy conclusions.

In spite of these two objections, embryology is of the very greatest assistance in enabling us to read the past history of animals, and to a less extent that of plants. Especially is this true of the early stages of the history. As we have above seen, the fossil records of animals become less and less satisfactory as we go farther back in history, and they finally cease altogether, long before we come to anything like a union of the converging types into a common starting-point. It happens, however, that it is of just these early periods that embryology gives us the clearest account. For various practical reasons, embryologists have largely confined themselves to the study of the early stages of the development. These stages teach of the relations of types to each other, and of the separation of the types from a common starting-point. In other words, they give us an idea of just that part of the history of life that we fail and shall forever fail to get from the study of fossils.

Closely allied to the study of embryology is that of anatomy. It has been determined that the embryological history of the higher animals of a class passes through stages which are represented by the adults of lower animals of the same class. This of course renders the study of lower types a valuable aid in tracing the history of the higher ones. It is also recognized to-day that relations between animals represent community of descent. When we find two animals closely related anatomically we interpret the fact as indicating a recent common point of origin. Anatomical relationship thus represents blood relationship, and if we have the former we can interpret the latter. It will, of course, follow that the study of anatomical relations will teach us history.

The study of anatomy and embryology cannot be isolated from each other. Together they give us an idea of the relations of types to each other, and enable us to draw up a picture of early development of life.

Miscellaneous Evidence.

But even when we carry the history back to the limits of embryological record, we fail to reach the beginning. Embryology starts with the living cell, and traces the growth of the organic world from this point. Can we learn anything of the origin of this cell? In thus going back to the very beginnings of the history of life, we leave fossils and embryos far behind, and have to look for our evidence elsewhere. There is no history yet discovered of the earliest stages of the living world. There is nothing known in nature which can tell when life first appeared in the world, or how. Upon this subject we can make only vague conjectures, instigated by various factors of nature. That in some way living things arose from that which was not living is certain, for at one time the world was unfitted from its molten condition for the existence of life. But whether life first came in accordance with the laws of nature which are still in operation, or in accordance with supernatural laws, is still a question in dispute, and perhaps may always remain so. All the evidence which can be brought to bear upon the subject is indirect, and can be comprised under three heads.

1. Experiments upon spontaneous generation, *i.e.*, experiments to determine whether life can to-day arise from the non-living. These experiments have been performed with a great deal of care and perseverance. They are, however, entirely incapable of touching the real question. A positive or negative conclusion would have little affected the great question under discussion. At the present time they

have been decided in the negative, but this only
shows that under certain conditions (which conditions were doubtless not those of the times when
life first appeared) life cannot arise spontaneously.
They tell nothing as to indefinite unknown conditions of the past.   If they had been decided in the
affirmative, they could show only that life could
arise under certain conditions, which conditions
again were unquestionably not those of early times.
The experiments have all started with a nutrient
solution, which contained already products which
were the result of life, *i.e.*, meat solutions, etc., and
such a condition would of course have been impossible at the time when life began.   At best, then, the
experiments which have been performed upon spontaneous generations could serve only to make us a
little better acquainted with life and the conditions
under which it now acts, without helping us a step
toward answering the question of its primitive origin.
It cannot be denied, however, that if this subject
had been or ever shall be decided in the affirmative,
it would render the origin of life by natural laws
more probable, since it would show that living things
could come from the non-living, and this would be
one step toward the solution.

   2. Study of organic and physiological chemistry.
It may seem somewhat strange to count chemistry
as a source of biological history, and of course the
evidence it gives is only indirect.   Chemistry has
been for some years now teaching us of the close
connection between chemical and biological laws.
It has shown that many organic bodies can be pro-

duced by purely chemical processes. It has shown
that many of the vital processes of living things are
simply chemical in their nature, and that some of
them may be imitated by lifeless material in the
laboratory. It has shown that chemical compounds
of great complexity have complex properties, and
that protoplasm is the most complex substance in
existence. In all of these ways has the close union
of chemical and biological laws been made evident;
and chemistry has thus prepared the way for the
conception of a still closer union, and for the supposition that life originally, as now, was a mere
application of natural chemical laws to complex
conditions, and thus arose by natural and not supernatural law.

3. The study of the low forms of life. This is of
value in showing the simplest conditions of life, and
therefore bringing us nearer to the condition of the
first life. The simplicity both of structure and
function of some of the lowest forms of life seems
to bring us very close to the inorganic world. The
step from the amœba to some inert chemical compound is certainly less than from man to the same
compound. From the study of these simple forms,
we can easily conceive of still simpler masses of protoplasm with even less organization than the amœba
and proteomixa. At the same time the complexity of
compounds manufactured by the chemist seems to
be approaching the somewhat greater complexity of
these simplest forms of life. It is plain enough that
the simpler the condition to which life can be
reduced and the smaller the gap between the sim-

plest living thing and the most complex compound
not alive, the less will be the difficulty in believing
in the natural origin of life. If we could find a
substance that was simply living matter with no
definite characteristics of any specific nature, this
would be the starting-point. From *a priori* grounds
we might expect that simple living matter would
appear before any definitely formed species. This
fact is the basis of the interest connected with the
study and speculation concerning protoplasm, the
supposed common factor of living things. In the
simplest forms of life we get down almost to pure
and simple protoplasm, and by taking from these
forms all of the common factors, we may suppose
that we obtain the characters of simple protoplasm
itself, and thus presumably the first form of life.

There is another series of facts which can hardly
be called evidence, but which does at the same time
have a great influence in our interpretations of the
past. The long-continued study of nature has led
to the formulation of the *law of continuity.* According to this law, the processes of nature have
been those of continual slow change, such that the
history of any minute is explained by the conditions
of the preceding minute. The law admits of no
great breaks in the history of the processes in nature,
but assumes that where such breaks seem to exist,
the break is only in our knowledge, and not in the
nature itself. It is impossible that the acceptance
of this law should fail to have great influence in the
interpretation of the life history, for by means of it
a constant development is necessarily substituted

for the series of epochs which the actual facts seem sometimes to indicate. Especially is this true of the conclusions as to the origin of life, for here, as we have seen, direct evidence is wanting, and the belief in the law of continuity forms the foundation of all that is to be said on the subject.

Such are the sources of the evidence from which our history of living nature must be drawn. Of its beginnings we know nothing beyond such inferences as may be drawn from experiments on spontaneous generation, organic chemistry, and the study of the lowest forms of life, together with the general teaching of the law of continuity. Once established, however, we can trace in outline at least the early history of animals and plants through the ages by means of the record we have of such history in their embryology and their anatomical relations. Later the stratified rocks begin to preserve for us here and there a scattered page of history ; and as we come to later ages, the leaves thus preserved become more and more perfect until in the latest times a fairly complete history may be read from them. Perhaps by the study of the course of the past we may then be able to hazard a prophecy as to the course of the life of the world in the future.

CHAPTER II.*

THE ORIGIN OF LIFE.

What is Life?

IN taking up the study of the history of life, we must first ask the question: what is life? This question is asked not in expectation that any satisfactory answer is possible, but in order that we may get as clearly as possible before our minds the chief facts in modern thought concerning the subject. It is clearly true that our scientists have by their speculations and experiments so completely changed our ideas of life that it sometimes seems as if we could almost grasp the real essence of the matter. The solution of the life question is said to be close at hand. ·It is well for us at the outset, then, to review the question, to see where we stand to-day in our knowledge of life, and to notice what points have been settled and what points still baffle comprehension. For, thus far, this life essence has been an *ignis fatuus;* and although many preliminary questions have been solved in pursuit of it, their solutions

---

* The substance of this chapter was originally published in the *New Princeton Review.*

only serve to show us that there is something else beyond, which is not comprehended. Our first task is then to find out exactly what is the question now at issue; for it is very different from what it used to be. We shall find that much is now universally conceded which was at one time strenuously disputed. There are three essential properties possessed by living things which must be included in any attempted explanation of life: *a.* Their constant activity. *b.* Their power of growth. *c.* Their power of reproducing themselves. These being the essential properties of life, their satisfactory explanation will bring us far toward the understanding of life itself.

Relations of Organic Activities to Physical Energy.

First we may say that the activities of organisms are no longer looked upon as manifestations of a distinct "vital force" unrelated to other forces. It will hardly be denied by any one to-day that all of the energy exhibited by organisms in their various activities is a part of the store of energy of the universe, and that all of the forces exhibited by animals are correlated with physical forces in general. It has been conclusively proved that every motion made by animals, every bit of heat arising in them, is simply a portion of the energy which this world has received from the sun. The process of its transformation is as follows:

Plants, by virtue of the possession of a body called chlorophyll (*i.e.,* their green coloring matter), have the power of using the energy of sunlight. By means of

2

the energy thus within their reach, they are able to
build complicated chemical substances out of such
simple compounds as water, carbonic acid, and simple
nitrogenous compounds. It is a well-known principle
of physics that to build a complicated chemical body
out of simple ones requires the exertion of energy,
just as it does to place a number of bricks one on top
of another.   All of the energy thus used is rendered
latent, but it can all be obtained again in active condition by pulling the structure to pieces. Every complex chemical compound may therefore be looked
upon as a store of energy. Plants, then, growing in
the sunlight are continually making use of the sun's
rays to enable them to build up complex compounds,
and they are therefore storing up the sun's energy
in the form of chemical energy. The energy of their
life, therefore, consists in transformed sunlight. Now
animals use as food the chemical compounds thus
built by plants.   Animals, unlike plants, are not able
to make use of the sun's rays directly, but they can
make use of the store of energy provided by the
plants.   They therefore derive all the energy of their
life by breaking to pieces these products of the
plant's constructive power. Just as the steam-engine,
by breaking to pieces the coal which forms its fuel,
makes use of the energy thus liberated, so the body
by similarly breaking to pieces its food makes use of
the energy thus liberated. The steam-engine converts the energy of chemical composition contained
in its coal into motion and heat; the body also converts the energy of chemical composition contained
in its foods into motion and heat. All of this is

practically granted everywhere, and we need not attempt to question the conclusion further. All of the energy of the body is a part of the physical energy of the universe, and its forces are correlated with other physical forces.

Relation of Vital Activities to Chemical Laws.

Again, it will hardly be questioned to-day that the chemical processes going on in the living body are fundamentally similar to those which may take place out of the body. The same laws of chemical affinity govern the changes taking place in the body and those occurring in experiments in the laboratory. The chemical processes of the body may be considered under two classes. The first are the processes of construction, by which simple bodies are built into complex ones. This class is chiefly found in plants. The second are those of destruction, by which the complex bodies are broken into simpler ones. This class is chiefly characteristic of animals, though found also in plants. The destructive changes are the simpler, and there is no reason to think they are any different from the destructive processes of the laboratory. The essential feature is oxidation, and oxidation may take place anywhere. It is true that the details of the process of this destruction in organic beings differs in some respects from that which chemists have been able to simulate. When food is thus broken up in organisms, many decomposition products arise which do not occur when the process is carried on in the laboratory. These products are thus characteristic of organic beings, and it

was for a long time believed that they could never
be obtained except through the influence of vitality.
Modern chemistry has demonstrated the possibility
of making many of them in the laboratory; many of
the simpler ones have already been manufactured,
and the list is constantly increasing.

Plainly, however, the destructive processes are by
no means so important as the constructive ones. In
plants the simplest compounds, H_2O, CO_2, and
NH_3 are built under the influence of sunlight, into
the most complicated ones. Even in animals this
constructive power is essential, for they do not con-
tent themselves with simply pulling to pieces the
products of the life of plants. They do destroy most
of them, but the energy liberated enables them to do
a certain amount of building for themselves. They
change dead matter into living matter, which must
be looked upon as a constructive process. Now our
chemists tell us that they have reason for believing
that even these constructive processes are purely
chemical, and will one day be simulated in the laboratory. They have, indeed, already shown that
many of these organic bodies can be manufactured
synthetically. Plants manufacture protoplasm, the
most complicated body of which we have any
knowledge. By the decomposition of this body may
be obtained a long series of decomposition products,
which become simpler and simpler until they are
once more resolved into the simple ones with which
the plant started. Now our chemists have begun
with these simple bodies, CO_2, H_2O, NH_3, etc.,
and have begun to climb this ladder of compounds

toward protoplasm. To be sure, they have not yet climbed very far, but they have made some advance. Many of the simpler members of the series have been manufactured synthetically from simple inorganic compounds. And since they have truly begun to ascend through this series, it is, of course, an easy inference to predict that they will some day reach the top and be able to make the higher members, even protoplasm itself. In theory they already tell us what the albumens are chemically, and expect to be able to make them synthetically before a great while; indeed, at the present time chemists are startled by the recent announcement of the manufacture of a proteid in the laboratory of Schutzenberger. Scientists are thus looking forward to the time when they will be able to make protoplasm in the laboratory, and thus artificially to make living things. Judging from the general tendency of advance, it does not perhaps seem improbable that they may some time make a body which shall have the chemical composition of protoplasm; but whether or not this body would be alive is the very question at issue.

Of course it is perfectly evident that the methods used by the chemist in these syntheses are very different from those employed by living cells. The chemist uses complicated apparatus and long, roundabout processes to produce the simple organic compounds. A transparent and seemingly structureless mass of protoplasm builds directly and with ease the most complicated bodies. No one will, of course, pretend to compare the organic cell with the chem-

ical laboratory in any exact sense. All that our
chemists can succeed in showing is that organic
changes are governed by the same chemical laws as
those which regulate inorganic changes. And the
possibility of the manufacture by synthetical processes of a number of the simpler organic compounds
gives us undeniable evidence that chemical laws are
the same in the body and in the laboratory.

Properties of Life as Explained by Physical and Chemical Laws.

Recognizing, then, that all the energy of organisms is derived from solar energy, and that the
chemical processes in the body are essentially similar to those outside, the next question to be
answered is how far the vital manifestations of
organisms can be explained according to these laws;
to see whether or not all of the activities of living
things can be explained by physical laws. And
here too when we reach the real opinion of various
thinkers, we find something like unanimity in many
points at least. Understanding the doctrine of the
conservation of energy, it is at once evident that all
of the energy displayed by organisms must be
transformed solar energy; and hence all of the manifestations of the body which are measurable by
units used in measuring other physical forces must
come under this head. Here will of course be included all the forms of motion, both molar and
molecular. The motions of the body, the heat of
the body, expansion and contraction of protoplasm,
all electrical phenomena, probably also nervous im-

pulses—all doubtless come in this category. Indeed, all manifestations of the body which can be matched by any machine will unhesitatingly be set down as coming under the head of transformed physical forces. We can believe that the body will do anything that a machine can do without calling in the aid of any distinct force. In other words, the activities of living things, though more complex, are as truly due to physical and chemical forces as those of a machine.

It is when we come to the other properties of life not found in machines that the problem becomes more difficult. No machine has the power to assimilate food and grow. These properties, which really are one, form the second character which universally distinguishes living matter from non-living matter. Now, so far as the mechanical process of growth is concerned, it is simply chemical change. This is certainly so in animals. They take into their body certain complex substances as food. This food undergoes chemical changes, chiefly those of oxidation. As a result, decomposition products are obtained, and some of these products of decomposition are ejected, while others are retained in the body, and thus the body grows. But it is not quite so simple as this, for, as we have seen, a certain part of the changes are constructive. Some of the decomposition products of the food are united together into more complex compounds, and all of it is more or less altered, so that none of the food is retained in the body in exactly the same condition in which it was taken. In short, the body assimilates its food, converting it

from its dead condition into its own substance. Now
it is not possible to imitate many of these changes as
yet in our laboratories, but that they are all chemical
changes can scarcely be questioned.   Even the constructive changes by which the body raises the
compounds into a plane of greater complexity, must
be regarded simply as chemical processes, the energy
which is required for them being obtained by the
breaking down of other portions of food. And in
plants also growth must be regarded as a chemical
process, for it consists in the combination of simple
organic compounds to form complex ones under the
influence of sunlight.

The third property of living matter—reproduction
—seems at first sight to be a more marvellous power
than that of growth ; but most biologists think that
it is easily derived from the latter. Fundamentally,
reproduction is a direct and necessary result of
growth.   In its simplest form, as found in the unicellular animals, it is seen to be nothing more than a
division.   The unicellular organism by chemical processes continues to assimilate food and thus to grow.
It keeps on increasing in size until it finally becomes
so large that the cohesion between its parts is in-
sufficient to keep the great bulk together, and as a
result it divides into two parts.   Each of these parts
is of course like the other, and there are thus two
organisms where there was only one before.   This is
the simplest case of reproduction, and heredity in
this case is to be explained as a necessary result of
growth.   Now, it is not difficult to see how the more
complex forms of reproduction may have been de-

rived from this. If instead of a single cell there were a large number of cells attached together, growth might lead them all to divide in a similar manner after the cohesion of parts had ceased to be sufficient to keep them together. And such a method of reproduction does occur in large groups of organisms. Or it might be that certain parts, perhaps a single cell only, would undergo this division, the parts of this cell becoming free to form new individuals, and thus spores would arise. Perhaps two individuals might fuse into one, and thus the vigor of both would be combined into one. This would lead to sexual reproduction. And so on. Of course the details of this process are purely hypothetical, and it is not our purpose to dwell upon them ; but it is easy to see that reproduction can probably be explained upon a mechanical basis as the result of assimilation and growth.

It is thus seen that the three properties of life can receive at least a provisional explanation from a mechanical standpoint in accordance with the laws of chemistry and physics, and since we are at present dealing with life in its simplest form, we need not here trouble ourselves with its higher properties of consciousness and intelligence. The explanations thus offered are accepted with practical universality by biologists, and we may regard the preliminary question in the problem of life as definitely settled.

Difference between the Dead and Living Organism.

With all of this explanation and reduction of vital manifestations to physical laws, no one can fail to

realize that something is lacking; and that though
scientists have explained much by hypothesis, they
have yet left the real question untouched.   It is not
easy to determine definitely this life factor,—a factor
so prominent in the minds of those who disbelieve in
the mechanical theory of life, and so readily ignored
by those who hold this theory.   That there is something more than has been reached by this explanation may perhaps be made evident by a further
consideration of the parallel between an organism
and a machine.   The comparison between the dead
body and the machine is exact, for each has the
mechanism which will enable it to transform one
sort of energy into another under the right conditions.   But in the body the requisite condition is
the presence of life, whatever that may be, which
guides the chemical changes taking place.   In the
machine the necessary condition is the presence of
an engineer, who guides the forces and chemical
changes.   The comparison of the living body should
not be simply with the machine in motion, but with
the machine plus the engineer.   This difference is
great indeed.   A machine may be ever so perfect,
and yet will not perform its work unless its engineer
supply its proper conditions.   Food out of the body
will never go through the complicated changes above
mentioned unless subjected to very peculiar conditions by the chemist.   Food in the body will not
go through these changes unless subjected to the
action of life.   Sunlight may fall upon CO_2, H_2O,
and NH_3 eternally without producing the slightest
tendency toward a synthesis of these elements.   But

let this occur in any living green plant, and how different the result. In some way living matter causes a synthesis to take place. The presence of life in an organism causes certain chemical changes to be set up in it which result in growth. Remembering that none of these changes will take place of their own accord, it is perfectly evident that there is something in the organism beyond simple chemical affinity,—some sort of power which directs chemical changes. Whatever it may be, it is the essence of life. In almost every sentence used in the comparison of animals with machines this factor can be seen. Even Huxley, the foremost in the mechanical theory, says, " We touch the spring of the word machine," and the result is speech, and the term " we" implies something not present in machines.

That there is a difference between organisms and machines at this point may be made more evident by consideration of the difference between living and dead organisms. That the body is a machine, and that like the machine it converts chemical energy into mechanical energy will to-day be everywhere admitted. But a machine cannot die. A machine may stop its motions, but a machine at rest is not comparable with the dead body. In both cases, it is true, there is a cessation of the changes which constitute activity, but in the one case the changes may be resumed again, in the other this is impossible. A dead body can never be revivified. It is more strictly to be compared to a machine which has lost its engineer, for with this loss disappears all possibility of further action. Its mechanism may be

perfect, and it may have all the possibilities of further action except a directing power, but without
this it is forever quiet. And so an organism is, so
far as we can see, frequently intact after death, with
all of its mechanism present; there is just as much
stored energy in a pound of fat in the dead body as
in the living body, ᴀ ɪd it is just as capable of being
oxidized. So far as we can see, therefore, every
physical condition may be present in the dead body
which is necessary to produce the process known as
life, if the process could once be started. But without this spark of life to start and direct the chemical
changes no life can show itself. And in like manner
do we find all other comparisons ever made between
organic and inorganic matter failing at this point.
Living things have been compared with crystals, for
both grow, although the process of growth is very
different in the two cases. But a crystal cannot die.
Take it out of the solution upon which its growth
depends and it will cease to grow, forever remaining
stationary. Put it back in the solution, and once
more it will resume its growth. A steel bar may be
magnetized, and under these conditions will exhibit
properties which it did not possess before. But it
may be demagnetized by a blow, and thus lose all of
these properties. This seems indeed to bear much
resemblance to death until we remember that a steel
bar may be magnetized and demagnetized indefinitely,
and never once fail to exhibit its properties. But an
organism, when it has once lost its vitality, càn never
be brought to assume a vital condition. It is perfectly
plain that at this point all of the comparisons of

organic with inorganic processes fail. There are some conditions supplied by the living organism not found in the inorganic world, and these conditions, whatever they may be, direct the play of chemical forces in the organism.

The Vitalistic Theory.

We have now finally reached the question at issue. The vitalistic question to-day is not to decide how many activities of organism can be explained by chemical and physical laws, but to discover what are the conditions which regulate these processes; to decide why it is that a living body can induce chemical changes which are impossible in the dead body.

One answer which has long been given to this question is that the necessary condition is the presence of a "vital force"; a force uncorrelated with other forces,—a distinct entity in itself. This force is life. It is conceived as having been supplied to the world at the beginning of life on the globe, and as having been handed down from one generation to another, or perhaps created anew at each birth. Vitality is therefore considered as something apart from the physical universe, but as capable of exerting an influence upon matter, to direct the changes taking place in it. According to this view, spontaneous generation would be an impossibility, for this vital force, not being derivable from other forces, could have its origin only from previously existing vital force. This theory labors under the disadvantage of being unable to say what is meant by

vital force, for of course we can get no conception
of any force except by its results.   But the vitalistic
theory claims that life is an immaterial something
which directs physical processes so as to produce the
activities which distinguish living things.   We need
not further consider this view, for it consists chiefly
in recognizing the necessity of something more than
chemical affinity and change, and in acknowledging
our inability to explain it by giving to it the name
vitality.

The Mechanical Theory of Life.

The point of dispute to-day is not whether the
vitalistic theory would explain facts, but whether it
is necessary.   A purely mechanical view of life has
slowly arisen from the profuse speculations, which
claims to be able to meet the case without recourse
to any imaginary "vital force."   The general tendency of scientific thought gives a certain amount
of *a priori* bias in favor of such a view.   It has
unquestionably been the tendency of science to
explain more and more of the phenomena of the
world in terms of the properties and laws of the
natural universe.   The foundation of the law of the
conservation of energy, the conception of forces as
modes of motions, are great steps in this direction.
In the organic world the theory of evolution, the
application of the conservation of energy to the
mechanics of life, the perception that the same
chemical laws govern living things and dead, and
every discovery of likeness between vital processes
and those purely mechanical, are all steps toward

this general unification. It is certainly in a line with this advance to reach a mechanical view of this life essence. If, therefore, a mechanical explanation is possible, there is good reason for believing that it is in the line of truth.

The mechanical theory is, in brief, that the directive conditions of which we are in search are simply those of chemical composition and molecular arrangement. It is pointed out that the *properties* of compounds increase in complexity with the increase of the complexity of the compounds, that as the molecule becomes more complicated its powers and possibilities become more diversified. The properties of compounds have, moreover, no traceable relation to the properties of the elements from which they are made. Oxygen and hydrogen, when they unite, form water, a compound with properties not possessed by either of the elements ; and yet we do not doubt that they are due to the properties of the elements. It is therefore easy to make the far-reaching assumption that,· when the molecule becomes as complicated as that of protoplasm, its properties will be as complicated as those of living things. One of these properties is to induce chemical changes in foods. Just as it is the property of water to dissolve many chemical substances, so it is the property of the highly complex body protoplasm to cause chemical changes. When it is possible, we are told, to manufacture the chemical substance protoplasm, it will of necessity be alive, for there are no peculiar powers in organisms not inherent in them as the result of molecular arrange-

ment.   The directive power which seems to exist is
no directive power at all, but only a property of
protoplasm.   Just as it is the property of platinum
sponge to cause, when held in a current of hydrogen, the hydrogen to unite with oxygen and burn,
so it is the property of protoplasm to cause more
complicated oxidations to take place, which produce
the fundamental process of growth, and from this,
as we have seen, other vital activities easily follow.
We see, therefore, that the comparison of the body
with the machine plus its engineer is replaced by
a machine that is purely automatic, and finds in its
own complex composition the conditions which
regulate its activities.   Death, according to this
idea, is simply the destruction of protoplasm, which
would, of course, destroy its properties.   Just as
soon as protoplasm begins to lose its complicated
structure, it loses all of the properties belonging to it
as protoplasm ; and this is death.   Demagnetism of
a bar of steel is therefore strictly comparable to
death, the only difference being that it is possible to
cause the steel bar to resume its former molecular
arrangement and once more to possess its magnetic
properties.   The possibility, however, does not exist
in living things; the violence of death ruins the
machine.   Even a machine cannot be started if its
adjustments are broken, and a living body being
more complex than any machine has its harmonious
action more readily ruined.   What is lost in death is,
therefore, not any directing force but chemical or
molecular composition.   The dead body is to be
compared not with a machine which has lost its

engineer, but with a broken machine which cannot be mended. Life is thus only an abstraction from the properties of living things, just as aquosity would be an abstraction from the properties of water.

This mechanical theory of life is not at present open to direct argument. The dynamics of protoplasm may be studied carefully; it may perhaps be shown that all the activities of protoplasm are easily explained as the result of chemical and physical forces. Already scientists are beginning to comprehend how the movements of protoplasm, which have proved so puzzling, are intelligible as the results of chemical change, whereby the density of the substance is altered, and consequently its shape. Indeed appearances seem to indicate that perhaps all the activities of protoplasm may be explained thus easily. But all of this fails to reach the real question at issue, which asks for the directive cause of these changes. The only direct argument would be to manufacture protoplasm and have it begin to assimilate food, or to show in some other way that a purely automatic machine is a possibility, which shall, as organisms do, supply itself with its own conditions of activity. Until this is done the mechanical theory can only be an inference from the general tendency of scientific advance.

The Origin of Life.

It is very plain that our verdict in regard to the origin of life in the world will depend largely upon what position we assume on the question just discussed. If we assume that life is a distinct force

3

unrelated to other forces of nature; if, in short, we accept the vitalistic standpoint, the matter becomes very simple. Such a force could not have come into existence except by a creative fiat. We should simply say then that far back in the history of the world some supernatural power introduced the first germ of life into the world, and that this first germ was the simplest form of protoplasm. To one who holds this view all the attempts to find a natural explanation of the origin of life are useless.

If, however, we are willing to accept, even provisionally, the view that life is not a distinct essence, but, as the mechanical theory of life would tell us, simply an abstraction from the complex properties of the substance protoplasm, then the question of its origin assumes a new aspect. It becomes then a legitimate question to ask how life arose in the world. Life by its inherent qualities is self-perpetuating, and if once it makes its appearance in the world its remaining here so long as the conditions admit is a matter of course. Geology tells us, however, that at one time the earth was so heated that no living thing could have existed on its surface. It follows from this fact, that life on the globe must have had a beginning. What then was the nature of the forces which brought the first living matter into existence?

Spontaneous Generation.

First we must ask if experiment or observation gives us any reason for believing that living matter can arise from that which is not living. Ever since

living matter has been studied it has been believed
by many that living organisms could arise spontaneously, *i.e.*, without having any direct living
ancestors. Aristotle held this view, and from his
time for centuries no one presumed to doubt that
most of the smaller organisms could, and usually did,
arise spontaneously. It was not until the sixteenth
century that the matter became one of discussion.
At that time Redi discovered that fly-maggots were
not produced spontaneously from decaying flesh as
had hitherto been believed, but came from something deposited by adult flies. This discovery led
him to further observation, and finally to the conclusion that there was no such thing as spontaneous
generation. Since that time this doctrine has been
the ground of many a hard-fought battle. The fol-
lowers of Redi speedily began to show by careful
study of the facts that numerous cases of so-called
spontaneous generation were simply due to careless
observation. It was soon proved that at least all
the higher animals arise by the method of reproduction only. The adherents of the belief that life can
arise from the non-living were thus driven to base
their claims upon the origin of the smaller organisms, and finally upon microscopic forms, which
can be studied only with extreme difficulty. But
so far from admitting this to be a retreat from their
position, they have shown that it is the most natural
conclusion possible. For *a priori* grounds should
serve to convince us that if living things can arise
from the non-living, this would be true only of the
very lowest organisms, those which approach the

nearest to the condition of simple protoplasm. The
disproof of the claims of the earliest biologists who
believed in the abiogenetic origin of the higher
animals, is therefore no proof or even indication
that this does not occur in the lowest organisms.
The question, therefore, finally settled around the
origin of the lowest and smallest forms of life.
From this point the matter has been chiefly one of
care in experimenting. It was found by some that
low organisms, bacteria, infusoria, etc., would arise
in closed flasks filled with various material for food,
even after all apparent precautions had been taken
to exclude everything alive. But other experimenters employing greater precautions for the exclusion
of living matter obtained opposite results. The verdict vibrated from one side to the other, as different
experiments were made known, until at length Pasteur and Tyndall showed that the negative conclusion was the only tenable one. Tyndall, more especially, by a series of careful experiments conducted
in a manner beyond reach of criticism, so conclusively proved that with the proper precautions no
living organisms could arise in any solution without
the access of previously living organisms, that no
one has seriously questioned the matter since. This
result, indeed, is only a negative one. It simply
shows that no living organisms did arise under the
conditions of the experiment. But it is so conclusive that scientists have, with practical unanimity,
given up all claim that there is the slightest evidence
for the possibility of spontaneous generation. And
this is admitted by the very men who still insist that

spontaneous generation must have occurred at some time in the history of the globe.

While it is thus true that scientists have somewhat reluctantly given up this fascinating theory, it by no means indicates that they have given up the belief in the possibility of life arising from the non-living under the right conditions. Although no one has as yet been able to produce conditions under which life can arise, this by no means proves that under different conditions a different result might not be reached. Protoplasm will not arise in closed flasks, but this does not show that it cannot do so at the bottom of the sea. If it could be shown that life arises spontaneously nowhere on the globe at the present time, this would by no means prove that in other ages, under different conditions, it may not so have arisen. And, indeed, now that the possibility of spontaneous generation to-day is practically decided in the negative, it is beginning to be recognized that the experiments thus far are utterly futile to settle the primary question at issue. Even if a positive result had been obtained, it would have had scarcely any bearing upon the question of the original appearance of life. This will be evident from the following considerations. The first living things must have been able to make use of inorganic material for food, since there could of course have been no organic food existing at that time. Our experimenters on spontaneous generation have, however, always used organic solutions in their experiments. Now, to-day only organisms which are supplied with chlorophyll are able to raise inorganic matter into an

organized condition. At the present time, at least, all
organic life depends upon the action of chlorophyll.
But in the experiments upon spontaneous generation
it has only been claimed that such organisms as bacteria and infusoria could arise spontaneously ; and
these organisms containing no chlorophyll have no
power to live upon the inorganic world. Our experimenters have found it necessary to supply them
with an abundance of organic food. Such organisms
certainly could not have been the first ones to appear
upon the earth, since they would be capable of existing only so long as organic food was supplied to them.
Indeed, if we could imagine the ocean filled with
albuminous food before any life appeared, and then
assume that these organisms could arise spontaneously, we should be no nearer to a permanent origin
of life than we were before. The only result would
be a rapid multiplication of these bacteria until the
ocean was filled with them ; the food would be con-
sumed, and then all would die of starvation, since
they would be unable to make food for themselves
out of the inorganic world as green plants can do.
The first living things must have been able to make
use of the inorganic world, and plainly, so long as
experiments deal only with chlorophylless organisms
arising in organic solutions, they have no direct relation to the question of the primary origin of life.

Should we then place the origin of life in the same
category of insolvable mysteries as the origin of the
universe in general? Looking at the universe in the
most extreme mechanical manner it is impossible to
think of it without some original creative power.

Behind the whole we must posit something which no thought can comprehend. If we must find creative power somewhere, perhaps the beginning of life may be an instance of its action. It may be well then, inasmuch as it seems probable that the origin of life can be nothing but a matter of speculation, to class it with the origin of matter and force, and thus to cease to explain it.

The question, then, stands something like this. There is not the slightest evidence to-day for believing that life can arise in any other way than through the influence of other living protoplasm. From the earliest living thing to the present there seems to have been a direct continuity of protoplasm, generation after generation resulting from the normal processes of reproduction, but in no other way. May it not be best to abandon the question of the origin of life and to say that it first appeared as the result of a creative fiat?

But this science refuses to do. Science grants that there are insolvable mysteries, and that the mechanical conceptions of the universe cannot explain all things. The origin of matter and force, the origin of motion and consciousness, are utterly insolvable mysteries, and are hence outside the realm of science. But it is thought that the origin of life is not one of the transcendent mysteries, but is one which will in due time be solved. This belief has been more especially prevalent among scientists since the precipitate advance of speculation in the last twenty-five years, due to the growth of the ideas comprised in the theory of evolution. This theory or group of theories has led to a belief in the general efficiency of natural law to account for natural phenomena ; and from this conception has arisen the claim that there must have been a natural origin of life. While then biologists have somewhat reluctantly given up their beliefs in the present possibility of spontaneous generation, many of them even the more strenuously assert that at some time, in some way, life must have arisen from the non-living.

Speculations as to a Mechanical Origin of Life.

Unable, therefore, to obtain direct evidence either for or against its proposition of a natural origin of life, science endeavors to meet the question by speculation. Having shown that vital processes are closely related to chemical and physical conditions, suggestions as to a possible causal connection between the two are of some significance. Speculations as to the origin of life can, therefore, hardly be called absurd, though they are almost unfounded in fact. Although they cannot be regarded as having much value, nevertheless modern scientific beliefs are in a measure founded on them.

We may pass over as irrelevant the suggestion that life may have been brought into the world by meteors. This does not of course assist in the slightest degree in solving the question of the origin of life. While different thinkers will hold different views upon the general question, nearly all will depend upon unknown conditions of the past for aid. A line of speculation something like the following would probably not be far from expressing the general outline of the thoughts of most biologists to-day who attempt to formulate any conception of the primal origin of life in accordance with natural law.

Assuming provisionally that life is simply a property of the complex composition of protoplasm, we can go on to ask ourselves how this complex composition could ever have been reached. Now certain facts of geology assist us much in this matter. During the early history of the globe the temperature was so high that few, if any, chemical compounds could exist. As the earth cooled by radiation, the elements hitherto kept apart began to come together in chemical union. All during the long process of cooling conditions existed which have never been matched since. Even after the temperature had reached a degree which admitted the existence of organic compounds, every circumstance was utterly different from what is found to-day. Different temperature, different relations of moisture, different electrical conditions, an atmosphere containing vastly more carbonic acid and oxygen than ours ; all these factors, and thousands of others of which it is needless to speculate, combined to make the conditions of chemical union widely different from any that can now occur. Under these circumstances it is plain that, with the universal chemical laws, chemical processes would be carried on of which we can know nothing, but which would be very different from any taking place in the world at present, or which can be simulated in the laboratory by the chemist or biologist. In these early times we thus see the possibility of production of

an almost infinite variety of compounds, each with
its own peculiar properties. Some of these compounds were so stable as to continue to exist down
to the present day, almost unchanged. Others were
constantly changing. The compounds of carbon
especially were varied and unstable, as we may conclude from the compounds of that element known
to-day. Many of these carbon compounds doubtless
would disappear with a change of conditions, breaking up, to enter into other combinations and form
other unstable compounds. Now amid this continued
succession of changes, the conditions of heat, elec-
tricity, etc., might at one time have been such as to
cause the elements, carbon, oxygen, hydrogen, and
nitrogen, all of which were present in the atmosphere,
to unite into certain complex bodies approximating
organic compounds. That this is a possibility becomes evident when we remember that our chemists
have already begun to make these elements unite
by laboratory methods. Many organic compounds
have been synthetically manufactured from inorganic material. Most of these compounds of early
times probably did not continue to exist very long,
since they were unstable, and had no power of self-
preservation.

' Thus far perhaps no one will hesitate to follow
the scientist, since he is dealing with authentic facts
rather than with speculation. But now he takes a
step into the dark. He supposes that at one time
these elements united into a compound which was,
owing to its peculiar composition, capable of causing
other bodies to change. By virtue of this power

other carbon compounds then existing were caused
to assume the composition of the new one, according
to the laws noticed in a previous section of this
chapter. Once this power is acquired the compound
possessing it would not disappear like the other
unstable compounds, but would remain permanent.
For this substance would assimilate food and grow,
and all the essential features of life can be deduced
from growth. This compound was of course protoplasm in its simplest form. It was only one of a
large number of complex compounds, which made
their appearance under the peculiar chemical conditions of early eras. Numerous others were doubtless formed, each possessing its own properties. But
only that compound which was capable of assimilation could continue to exist in an active condition
during the subsequent ages. This substance eventually absorbed all other compounds in any way
similar to itself which may have arisen contemporaneously with or before it, and it remains, therefore, to-day the only living matter, the physical
basis of life.

It will be seen that, according to this speculation,
the first form of living matter was by no means
similar to any organism of to-day. It was rather a
diffused mass of protoplasmic substance, with no
differentiation into cells or parts or individuals. It
will be further seen that it is not necessary to assume that this first protoplasm possessed chlorophyll
as has been claimed, for, according to the hypothesis,
there were many other carbon compounds of high
complexity produced at the same time. These com-

pounds, more or less similar to protoplasm, though
not capable of self-perpetuation, would serve the
first protoplasm as food.   There would thus be no
lack of organic material for the subsistence of life of
the first protoplasm, even though this first organism
were incapable of feeding upon the inorganic world
directly.   Doubtless, too, the conditions which produced the first living protoplasm existed for a long
time, and thus living matter would for a long time
be brought into existence by processes other than
those of reproduction.   Indeed, there was no definite beginning of life.   Here, as elsewhere, nature
made no jump, but produced life as she produces
everything else, by slow stages.   Chemical processes
of early times resulted in the production of many
compounds which, acting upon each other, and acted
upon by the changing conditions, became modified
in an infinite variety of ways.   Their complexity
and instability became very great.   Finally, some of
the most unstable of all began to effect changes in
others which resulted in assimilation, and thus slowly
the properties became more marked.   Simpler and
simpler substances were made use of as food.   So
long as the original conditions lasted there would of
course be no need that living matter should possess
the properties of chlorophyll.   Nor was this at all
necessary while circumstances were such as to make
possible the natural development of high carbon
compounds.   Eventually the power to live upon the
simpler inorganic foods must have been acquired.
But it is only necessary to assume that this power
became fully developed by the time that the con-

ditions had so changed that protoplasm could no longer be developed by the original spontaneous method. Perhaps for ages protoplasm existed unable to use organic food, but finding sufficient food in the surrounding complex carbon compounds. And when this power did at last become developed, it was not acquired by all protoplasm. For just at this point the organic world became divided into two parts. One part did develop chlorophyll, and has since been able to live upon inorganic matter, using the energy of sunlight to build this matter into an organic compound. Finding its food, carbonic acid, water and nitrates and sunlight everywhere, this class of organisms did not acquire the power of motion. The other half of the living world never developing chlorophyll became of necessity at last parasitic upon the plants, and developed an almost universal power of motion in order to enable it to seek food. The animal and vegetable kingdoms were thus finally separated from each other, with the relations which they hold to-day.

Such, in brief outline, is the substance of some of the modern speculations concerning the method by which life arose. It represents one phase of such speculations, and is subject to great modification in the minds of different thinkers. It is plainly open to sufficient criticism, and it is equally clear that it is not capable of direct proof, at least in the present state of science. It is as moderate in its terms as any of the suggestions upon the subject, and makes as slight claims upon our credulity. It will, at all events, serve our present purpose of giving an idea of the relation of speculation to the question of the origin of life.

The Significance of such Speculations.

Now what are these speculations worth? Many will immediately answer that they are worth nothing. Others may regard them as having a certain amount of suggestiveness, but no great value. It is perfectly plain to every one that they are purely hypothetical. Not only are they unproven hypotheses, but they are further of such a nature that there can be no evidence either for or against them. They must unhesitatingly be set down as scarcely more than bold guesses at a possibility. Even Huxley says: "Of the causes which have led to the origination of living matter it may be said we know almost nothing." If, then, science is to confine itself to facts, these suggestions may be cast aside as worthless. Why is it then that we find so many biologists to-day willing, yes, more than willing, anxious, to accept them? Certainly it is not because they are the simplest explanations, not because a large number of converging lines of thought point toward them. Those who seriously discuss these speculations, or regard them as of any significance, do so from some cause lying outside of the question itself.

And this cause is to be found in certain philosophical conceptions. Science studies the world from one standpoint only; a standpoint which its devotees naturally believe will lead them most surely to the truth. This study of nature from the exterior has led to the grand generalization that all nature is governed by law. The significance of the word law does not particularly concern science, but is left to other realms of thought. Science satisfies itself in discovering and applying laws. A thorough study of nature has made it seem probable that natural law, when thoroughly comprehended, will explain all natural phenomena. So many facts formerly relegated to the realm of the supernatural have been explained by natural law that science has determined to call in the supernatural as seldom as possible, and to accept no breaks in the chain of law unless absolutely forced to do so. This generalization is at the foundation of the terms—law of continuity and evolution, as they are used by science to-day. The significance which this question of the origin of life has for all evolutionary theories is at once evident. It is an important link in the chain of continuity, for unless the spontaneous generation of life be a fact, the law of continuity is no law. For even if science does succeed in explaining the development of life from the lowest to the highest, but does not explain the origin of this first form, it has only half accomplished that for which it is striving—viz.: to reduce living phenomena to the same laws which govern the non-living. It is not surprising, therefore, that we find biologists observing, experimenting, and speculating, in order to find some way to help themselves out of their dilemma. To any one who is inclined to believe in this law of continuity, and the efficiency of natural forces, such speculations as the above, which show a possible avoidance of a break at the beginning of life, have a

certain amount of significance. If we are inclined to
believe that "nature does not make jumps," it follows
that every break which we see in the continuity is not
a break in reality, but simply in our knowledge of
history. Many breaks which formerly existed in our
knowledge have disappeared with advancing discovery. It is natural, then, to believe that the
present chasm between life and non-life was, at the
beginning of the world no chasm, but filled with lost
stages which can never be recovered. Speculations
as to the nature of these lost stages have, therefore,
some meaning in the light of the law of continuity.
Scientists do not look upon any of them as necessarily or even probably true. They do not consider
that we have sufficient knowledge to say anything
definite upon the subject. But science does look upon
these speculations as indicating that the problem of
the origin of life is not an insolvable one. Scientists
take them for what they are—pure speculations, but
think that they tell us that the break at the beginning
of life is one of ignorance and not one of fact.

It is thus only the supposed existence of a philosophical necessity which has created a demand for
some theory of a natural origin of life, and called
into existence the various speculations on the subject.
The conclusion has been reached that the general
advance of thought and investigation has practically
established the truth of the law of continuity. This
law, so thoroughly believed in by modern science,
demands the destruction of the chasm between the
living and the non-living. Science has, therefore, set
to work to destroy it. It has shown that the chasm

is not so great as was once thought ; it has proved
that the animal body, and protoplasm in general, is
a machine making use of the chemical energy of its
foods ; it has shown that growth is little more than
chemical change, and that throughout the organic
world the same physical and chemical forces are at
play as in the inorganic world, only under more complex conditions ; and it has rendered it probable that
most if not all of the vital properties are directly
dependent upon and explained by chemical and
physical forces. Science has, in short, proved that
living processes are a continuous change of chemical
and physical forces, and that what we mean by life is
something to direct this play of force. It then
assumes that this something is to be accounted for as
the property of protoplasm resulting from its complex composition. This assumption is plainly a long
step in the region of hypothesis. But once made, it
becomes easy to posit and to explain by speculation
the spontaneous origin of life. For, indeed, it now
follows as a matter of necessity. The conclusion
which experiment forces upon us, that spontaneous
generation does not occur in nature to-day, is cast
aside as irrelevant to the more fundamental question.
For we ought not to expect, even if life originally
did appear mechanically, that it could do so now,
since the conditions are so different. Concerning the
first origin of life, science, therefore, knows nothing,
and is obliged to rest satisfied with the statement
that its original mechanical origin is an absolute
necessity of thought. "To hold the beginning of
life as an arbitrary creation is to break with the

4

whole theory of cognition," says Zöllner. To the scientist who is convinced of the universal truth of the law of continuity, therefore, the natural origin of life, though not possible now, was possible and did occur in early times under conditions about which we can only speculate. Carbon in former times certainly did crystallize in the form of diamond, because of conditions which then existed, and it does not do so now because of the absence of those conditions. So, we are told, the elements carbon, oxygen, hydrogen, and nitrogen, did in former times unite together to form protoplasm, under conditions which then existed, but have long since passed away.

But this may seem to be attacking the problem from the wrong end. The law of continuity is the law to be proved. To any one who is disposed to question the far-reaching significance of this law the matter is by no means so evident. If we are willing to accept the existence of breaks, few or many, in the history of the universe, we may well place one at the beginning of life. The break exists to-day, at least, and no amount of ingenious speculation is sufficient to cover it. Incapable of proof or disproof, demanded by no bit of evidence, and if we do not accept the law of continuity, not demanded by any philosophical necessity, these speculations of science will be regarded as worthless. They are laborious searches after something which does not exist. But on the other hand, when we remember how persistently scientific advance is leading us toward a mechanical conception of the phenomena of nature, and how strongly the law of continuity is establishing itself in the light of discovered fact, we must admit that there is a significance in speculations on the origin of life which is deeper than at first sight appears.

Protoplasm Not the Simplest Basis of Life.

In the discussion which has gone before, protoplasm has been referred to as if it were a definite chemical compound, though one of great complexity. It has seemed to many that the first step in the history of life must have been the production of a protoplasmic mass like some of our present low organisms (*Myxomycetes*). But nothing is more certain than that protoplasm is not a definite substance, and that it is by no means uniform in different organisms. It is found to differ widely in its physical nature, sometimes being almost solid, sometimes very liquid. It is found to differ more or less in its chemical composition, no two analyses agreeing exactly, and when we take its functions into consideration, the differences between different specimens of protoplasms become extreme. Although the amœba and the nerve cell of the human brain are both composed of protoplasm, how wide are the differences between the powers they possess. Plant protoplasm is capable of making use of the energy of sunlight, while animal protoplasm is not. When we call protoplasm " the physical basis of life," and try to prove the unity of the organic world by its universal presence in all living things, we have not by any means reached the bottom of the matter.

We have, it is true, taken one step farther toward simplifying the life problem when we step from the study of the cell to the study of protoplasm, but we have not reduced life to a common factor. There is, indeed, no such thing in nature as protoplasm in general, but only particular kinds of protoplasm. Protoplasm can produce new protoplasm, but not new protoplasm in general, only new protoplasm just like itself. The amœba can make new amœba protoplasm and no other, while the nerve cell is capable of giving rise to new nerve protoplasm. The study of the variations among protoplasm has progressed until we are slowly ceasing to look upon it as a definite substance. The microscope investigation of recent years has shown that not only is it chemically complex, but it can no longer be regarded as physically homogeneous. Several distinct chemical compounds have been discovered in it. Within the simplest protoplasm has been found an extremely complex structure. There is a dense reticulum made up perhaps of hollow threads, containing minute granules and a liquid matter (Fig. 1). The granules are united in various ways to form groups. To-day we may almost say that the study of the structure of protoplasm promises to become as complex a subject for the future as the study of the animal kingdom on a large scale has proved to be in the past. Instead of looking upon protoplasm as the primitive life substance, at the present time, most special students of protoplasm would regard the granules as more likely the fundamental part and all else derived. These granules are excessively

minute, requiring the highest powers of the microscope (four thousand diameters) to see them at all,
and yet they are now looked upon as having in
themselves all the remarkable powers of life. We
may perhaps even say that wherever there is a single

Fig. 1.   Protoplasm, showing its reticulum, very highly magnified.

granule there is life. By the activities of the granules growth takes place, by their differentiation and
union and by the products of their growth is formed
the compound which we have called protoplasm.
By their aggregation into definite groups are formed
the cells. The differences between cells are due to
the differences in their arrangements. Protoplasm

of itself is thus ceasing to play so important a part
in biological discussion as it promised to do ; its
place as a primitive life substance is being taken by
the granules with their various products, deuter-
plasm, microsomata, etc., etc., and in scientific litera-
ture to-day we hear less and less of protoplasm.
Protoplasm is no longer the physical basis of life, but
if indeed scientists should to-day recognize anything
under this term, it would probably be the granules
which are found in such abundance within the protoplasm.   And yet even these minute bodies, small as
they are, can hardly be said to have reached their
ultimate analysis.   They show different functions,
in different cases, and cannot be looked upon as
alike.   Whether the analysis can ever be carried
farther, of course we cannot say.   But certain it is
that protoplasm cannot be any longer strictly regarded as the physical basis of life.   To-day the
only universal properties of protoplasm seem to be
that it is reticulated and has certain powers of motion, growth, and reproduction.   It is thus seen to
be always a secondary rather than a primitive substance, and the term protoplasm seems to be an
abstraction from the idea of many organisms, just
as the term mankind is an abstraction from many
individual men.

There are therefore many steps between the simplest protoplasm yet discovered and the inorganic
world, or even between it and the simplest form of
living matter.   Hence it is impossible to look at a uni-
formly diffused mass of protoplasm as the first step in
the life history.   But however this may be, it will be
seen that it does not materially alter the significance

of the scientific speculation of the present chapter. In showing the complex nature of protoplasm, the microscope has perhaps given a little more foundation to the supposition that its properties are due to its complex composition, physical as well as molecular, but has not of course removed the difficulty of a perpetually automatic machine, acting without any directing power outside of itself, and has made its mechanical origin a little more difficult of comprehension. The substitution of these granules for protoplasm, moreover, only throws the question of the origin of life a little farther back into the microscopic field, but does not alter the problem as to the forces which brought it first into existence.

Life Appeared in the Ocean.

Although, as seen above, the beginnings of life are shrouded in darkness, there is one point that can be stated with certainty. There is no doubt that the first forms of life appeared in the ocean. Protoplasm itself is largely water, and in the water it must have appeared. At the early periods, when life appeared, there was probably little or no dry land, and there could have been no bodies of fresh water to serve as the starting-point of life. Add to this the fact that all of the earliest forms of life found as fossils are marine, and it becomes certain that the ocean must have been the place where the first living thing made its appearance.

Summary.

Life, as we commonly use the term, is certainly an abstraction from the properties of living things.

Whether in its ultimate analysis it is still to be so
regarded is an open question. The marvellous powers of living things, and the great mystery surrounding the processes of growth and reproduction, have
caused thinkers to believe that the property of life
was something more than a simple abstraction, and
to regard it as.a distinct and mysterious force resid-
ing in organisms, capable of indefinite expansion, and
having its origin in some direct creative power.
This conception has been retained while the belief
in the general mechanism of nature has been gaining
universal acceptance.

This conclusion, that of the vitalistic school, has
for some time been subject to criticism. The mysterious activities and powers of living things have
been in a measure explained as the results of known
forces of nature. The discovery of the law of conservation of energy, the discovery of the close relation of chemistry to vital processes, the manufacture
of organic products in the chemist's laboratory, and
the study of the various properties of the complex
compounds of carbon have all united in proclaiming
a close relation between the laws governing the living and the non-living world. It was inevitable that
the conclusion would be reached that what we call
life is nothing more than a manifestation of ordinary
forces of nature under special conditions furnished
by the substance protoplasm. Such a position is
assumed by the mechanical theory of life.

Even when this position is taken, the fact remains
that at the present time protoplasm does not arise
except through the agency of other protoplasm.

We are thus forced to ask what could have been the origin of the first protoplasm from which the rest has arisen. After all the discussion, however, we must finally admit that we do not know when life first arose or how, nor do we understand the causes which brought it into existence. Many secondary problems have been and are being solved, but the real question remains as yet untouched, except by hypothesis and speculation. Vital processes may all be shown to be chemical and physical processes, but this will never explain why they are carried on automatically in protoplasm and here alone; and granting, if we are inclined to do so, that it is one of the physical properties of protoplasm to direct this play of force, there still remains the fact that to-day protoplasm can only come from other protoplasm. Whence came the first protoplasm? To this question science offers in answer, first: the law of continuity, in terms of which the original spontaneous generation of life from the non-living is a necessity; and, second, various speculations, which, though acknowledged to be entirely unproved, are regarded as showing that in the boundless possibilities of the past, spontaneous generation might well have taken place, provided always it be granted that life is simply the result of complex chemical and molecular composition.

From all of this discussion and speculation we can isolate two facts: 1. Life arose in the ocean. 2. The first form of life was the simplest possible condition of living matter, certainly simpler than any living organisms with which we are acquainted to-day, and

very likely simpler than the simplest mass of diffused
protoplasm.   Amid the confusing and unsatisfactory
mass of fact and speculation which centres around
the primal origin of life these two points may stand
prominently before us as certain, while all the rest is
hypothesis, with various degrees of probability.

CHAPTER III.

THE ORIGIN OF THE ANIMAL KINGDOM.

WE have now the history of life fairly started. Living matter made its appearance in the geological seas of the far distant ages, and was at first probably an undifferentiated mass with the simplest powers of assimilation and growth. Out of this the higher orders of life were to arise, and we have now to study the first steps toward the modern world.

Unicellular Life.

The first trace of anything leading toward the modern world, of which we find evidence, is in the formation of cells. It has been suggested in the last chapter that there first existed a diffused mass of living matter without differentiation into parts. Such a conclusion has not been proved, and if such a mass did exist, it probably continued only a short time. It soon became broken up into small masses which were more or less independent of each other. Our reasons for this conclusion are these: The simplest and lowest organisms which exist to-day consist of just such little independent bits of protoplasm which we call cells (unicellular organisms).

Secondly, from the study of embryology we find that all animals and all plants begin with their development as single cells, and since embryology repeats past history this of course points to a former unicellular condition of early animals. Here, then, is our first sure starting-point. For a diffused mass of protoplasm at the beginning of life we have only a speculative foundation, but for a unicellular condition of early life we have direct evidence.

Cells.

Before tracing the history farther, it is well therefore to say a word in regard to the meaning of the term cell, as it is understood to-day by naturalists. As originally used, the word cell was a good description of the object so named. The microscopic study of plants showed long ago that they are in all cases composed of a larger or smaller number of little separate compartments or little boxes, like a mass of honeycomb. Each box has a wall and contains a mass of the living substance, the protoplasm. These were named cells. With such objects as a starting-point, for which the term cell was a good description, the word has been extended in its application until it has come to be applied to a very different set of objects. It was soon found that all plants and all animals were composed of somewhat similar independent parts and the name cell was naturally extended to animals as well as plants. Inside of these cells there was subsequently shown to be frequently present a bit of dense protoplasm quite distinct from the rest, which received the name of nucleus. In many cases, however, especially among animals, the independent bits of protoplasm were found to have no wall to surround them, and many cells were found in which no nucleus could be demonstrated. Still they were seen plainly to be

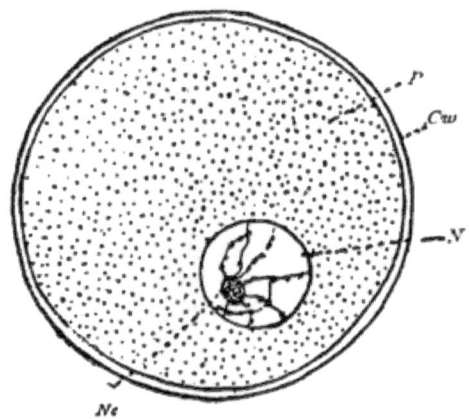

FIG. 2. A cell.—*P* Protoplasm. *N* Nucleus. *Ne* Nucleolus. *Cw* Cell wall.

independent units and surely to deserve the name of cells. Thus the term cell came to mean any independent mass of protoplasm, with or without a nucleus, with or without a cell wall. More recently however, it has been quite definitely shown that, though the cell wall is not always present, and therefore is not necessary to the constitution of a cell, the nucleus is probably always present in active cells, either as a distinct body or as a diffused mass. A cell thus becomes an independent bit of protoplasm with a nucleus; or better, perhaps, since recent study shows the fundamental importance of the nucleus, a bit of nuclear matter with surrounding protoplasm.

Fig. 2 shows a cell in the modern sense. All of the
living parts of animals and plants are composed of a
mass of such cells. The cells however, are far from
being all alike. Indeed, they differ extremely in
shape and size and in many other respects, but in all
cases each consists of a bit of nuclear matter sur-
rounded by protoplasm, and each is thus more or
less independent of the others. The larger and
higher animals are composed of a larger number of
cells, and always of a larger variety in their shape
and function.

The cell has thus been for many years regarded as
the unit of life, and the life of the individual as the
sum of the life of its cells. When we go among the
lowest and simplest of animals, we find one large
group in which the individual consists of only a
single cell. These are the unicellular animals (*Protozoa*) and plants (*Protophyta*). They are all aquatic
organisms, all small, usually microscopic, all of course
extremely simple, but withal extremely abundant :
they are found in the ocean, in rivers, brooks, ponds,
pools, ditches, swamps, gutters, puddles ; in short,
wherever there is any water, there we may be pretty
sure to find some representatives of the unicellular
organisms. Fig. 3 represents a few types of these
animals.

It was a great simplification of our knowledge of
animals and plants when it was discovered that all
were to be regarded as complexes of these simple
cells, for it gave a unit of life. The unicellular animals
and plants were naturally regarded as the simplest
possible organisms, since they consisted of one such

unit. It was therefore extremely significant to find
that in their embryology all animals and plants alike

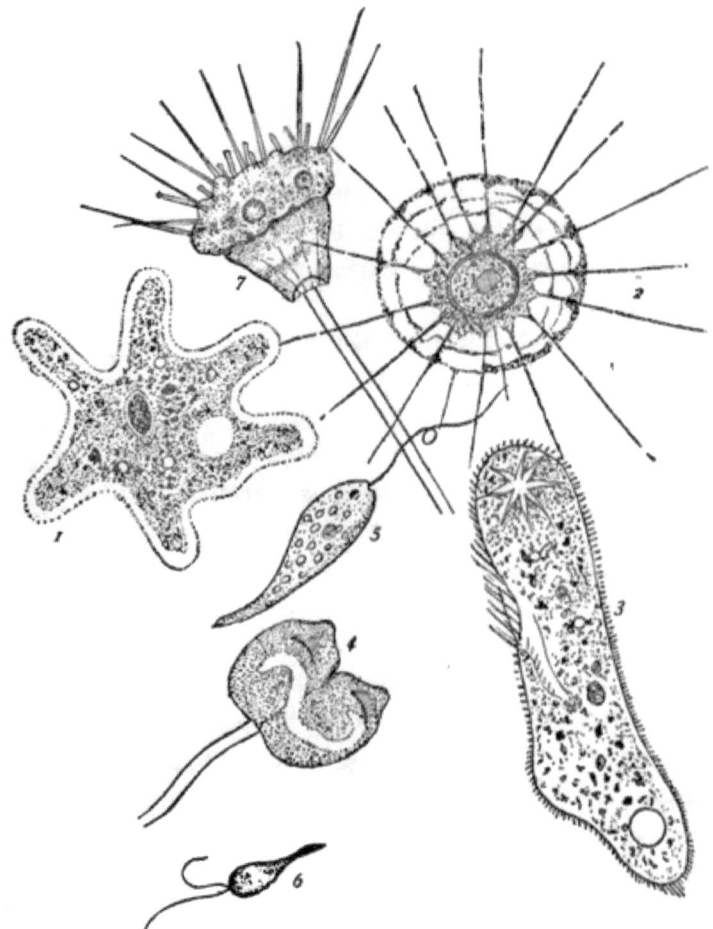

FIG. 3. Types of Protozoa.—*1* Amœba. *2* Actinospherum. *3* Paramecium.
4 Vorticella dividing. *5* Euglena. *6* Trichomonas. *7* Podophyra.

began life as a single cell, the egg and the germ cell
of the plant being always single cells. To find the
cell as the unit of life, to find a group of unicellular

organisms occupying the lowest position in the scale
of nature, and to find all animals and plants beginning their embryology as single cells, were coinci-
dences of remarkable interest.   There can be only
one interpretation of this.   Since embryology is an
epitomized account of past history, the fact that all
animals and plants begin life as single cells, of course
must mean that, if we could follow back the history
of animals and plants, we should in each case finally
come to some unicellular organism.   A unicellular
organism was therefore a common starting-point of
all animals and plants, and the Protozoa and Protophyta of to-day are of interest as being close representatives of the earliest organisms of which we have
any suggestion in our recorded history.

This conclusion of a unicellular starting-point of
all animals and plants is a significant result of biological study.   At the same time, as was seen in the
last chapter, more recent and exhaustive study of
cells is beginning to show that the simple cell is not
itself the unit of life, but is a complex body.   Processes are found to take place inside the cell which
indicate that we are still far from the unit of life.   It
is seen that of the whole cell the nucleus is really the
essential part, and that it regulates the activities of
the rest (Fig. 2).   The nucleus itself is moreover a
complex.   No fewer than five different chemical
compounds have been found to exist in it.   In structure also we find various fibres, liquids, and granules,
all of which seem to undergo definite cycles of change
in the activities of the cell.   All of this indicates that
we are still far from the unit of life when we have

reached the single cell. There is, perhaps, as broad a series of phenomena to be studied between the single cell and the real unit of life, as those which we have found between the higher animals composed of many cells and the single cell. Animals have all been reduced to complexes of cells, but the cell bids fair to be still farther reduced to a complex of inconceivably small granules. Be that as it may, it still remains a fact that in the history of life we have not yet discovered any direct evidence of an earlier condition than that of the unicellular organisms. Embryology, which is our first source of evidence, gives us a record of a unicellular condition, but of nothing earlier than this. The first step that we can take in narrating the history of life is, therefore, to establish the period when unicellular animals were the highest condition of organic life in existence. The diffused protoplasmic mother of life, if ever such existed, had become broken into independent masses.

Of course, nothing can be determined in regard to the variety of form of these early organisms, nor can we say whether they were all alike or whether they showed variety like that of the unicellular organisms to-day. Of their habits we know nothing. That each organism was independent and capable of carrying on all the functions of life within itself follows from every source of belief. That these organisms multiplied by simple division is certain from the following reasons: Multiplication by division is a universal phenomenon in all living things to-day. In its simplest form it is shown by most of the existing unicellular animals, as is illustrated in Fig. 4. As there shown, multiplication simply consists of the division of the original cell into parts, each of which is like the other, and each of which is henceforth independent of the others. Now such multiplication is universal in the unicellular organisms; such a multiplication is nearly universal among the cells

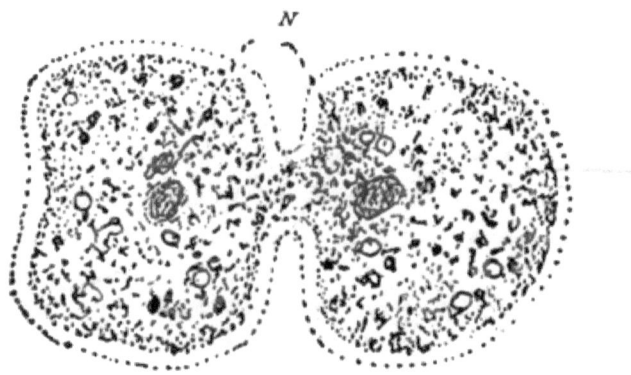

Fig. 4. Amœba in the act of dividing.—The nucleus, *N*, has already separated in two parts.

of the bodies of higher animals; such a multiplication is the universal method by which the single-celled ovum begins to grow and develop into the many-celled adult. From all of this there can be no room for doubt that such a method of multiplication was possessed by the earliest unicellular animals, which we must assume lived at the very beginning of the history of life in early ages.

The history of the living world has been like that of a branching tree, a main trunk dividing into branches, and these in turn subdividing until they become lost in thousands of small twigs. From the time of branching it is of course impossible to follow the history as a whole, and only possible to follow that of the various branches. At the point which we have now reached we have come to the end of the trunk, and we find the first division into two large branches, animals and plants, as mentioned in the preceding chapter. The separation between these two kingdoms seems to have occurred even while the living world consisted of unicellular organisms. From this point we must follow the animals and plants separately. We shall first take up the history of animals, leaving that of plants for later consideration.

Early History of the Animal Kingdom.—The Origin of Multicellular Animals.

The Protozoa as they exist to-day may be fairly supposed to illustrate the early unicellular animals of pre-historic times. The next stage in the history of life was the appearance of the multicellular organisms. It is again the study of the unicellular animals to-day which gives us suggestion as to how this change arose. Among most of the unicellular organisms, when a cell divides, the parts completely separate from each other and live subsequently independent lives. Among a few living forms, however, the parts do not completely separate from each other, but remain together in some sort of connection and are more or less dependent upon each other. Fig. 5 shows such a group of cells, the members of which have arisen from the division of a single one, and which

have not completely separated.   There is a small
amount of dependence of the cells upon each other;

FIG. 5. A Colony of Protozoa.—Each cell is independent. *a* and *b* show manner
of division to form colonies.

they share each other's food, and all seem to be connected by some physiological bond.   Still, each is in
itself complete individual, having a complete system of

organs, and each can live by itself perfectly well if separated from the others. Such a group does not form a multicellular animal, since the cells are each complete individuals. It is rather a colony of unicellular animals. Nevertheless, such a colony is the first step toward the production of the multicellular organisms. From embryological evidence also there can be little doubt that such was the method by which the multicellular animals and plants began to develop in the early history of the world. We have no means of determining to what extent this formation of such colonies of independent animals occurred. It was indeed only a stepping-stone toward the next stage in the history of living things, a stage of much more importance, viz., the formation of the first true multicellular organism.

To form a multicellular organism it is not sufficient that there should simply be a large number of cells attached together. This occurs in many of the animals that are regarded as unicellular. In order that there should be a true multicellular animal, a metazoan in distinction from the protozoan, it is necessary that there should be a certain amount of division of labor among the cells. So long as each cell carries on all the functions of life in itself the aggregation of cells forms simply a colony, but when the different cells in such a group begin to assume different functions—*e. g.*, some of them capturing food and others digesting it,—then the different cells become strictly dependent upon each other, and there arises a true multicellular animal. A multicellular animal is one in which there is an aggregation of cells, each performing only a part of the functions of life, and thus each dependent upon the others. The whole forms a unit. In such a community it is no longer possible for a single cell to be separated from the rest and still continue its life, for such a cell would be able to perform only those duties for which it was adapted, and it would therefore soon die.

The Gastræa, the Common Trunk of the Animal Kingdom.

That such a multicellular community arose in the early history of life from the unicellular forms, is of course evident, and that it arose by cell division which did not become complete enough to separate the individual cells from each other, is also almost certain. Exactly how a division of labor first arose, or what that first division may have been, it is perhaps impossible to say. Indeed, it is not improbable that in the early history of life there may have arisen many different types of division of labor giving rise to different kinds of true multicellular communities. But whatever may have been the early varieties of such differentiation, the very important fact is true that only one of these early types of multicellular animals perpetuated itself so as to affect the subsequent development of animals. The study of embryology shows that all of the subsequent types of the animal kingdom arose from one definite type of early differentiation. In other words, of all the types of cell aggregates or communities which may have been produced in the early history of life only one of them proved itself of value enough to take a definite place in the history of animals, and from this one all other types of animals have developed.

The early type of animal thus referred to as of so much importance in the history of living things is one of which we have abundant evidence in embryology, though perhaps the type does not exist to-day as a distinct animal. Since embryology is the only

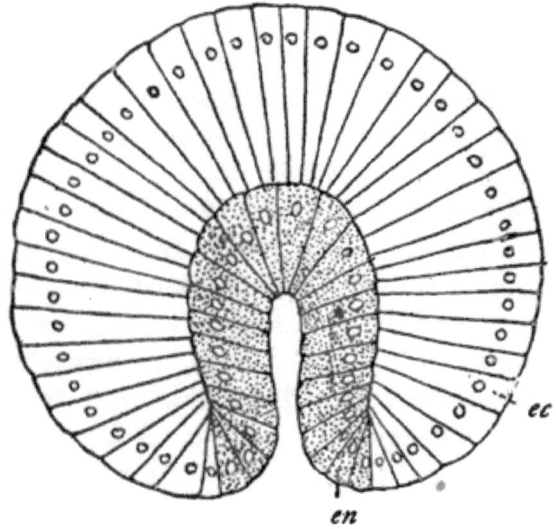

FIG. 6. A Typical Gastrula.—The shaded cells, *en*, are those connected with the digestion of food. *en* Endoderm. *ec* Ectoderm.

evidence of this stage in the history of life, our knowledge of this early form is, for reasons shown, confined to its fundamental structure with practically no sure knowledge of its details. The type in question has been named the Gastræa by Haeckel, who first clearly saw that embryology taught that such a common ancestor formerly existed. The evidence for this conclusion is as follows: It has been found by a

long-continued study of embryology that all multicellular animals pass through a stage which embryologists have called a gastrula.   A typical gastrula is
shown in Fig. 6.   It is seen to consist of a cup made
of two layers of cells, one of which forms the outside
of the cup and the other the inside.   When the two
layers are thus formed they assume different functions; the outer layer of cells becomes especially
connected with the powers of motion, sensation, and
the other functions directly connected with the outer
world.   The inner cells being removed from direct
contact with the exterior, and being in a measure
protected, take for their function the duty of the
digestion of food which is supplied to them by the
outer layer of cells, the opening serving for both
mouth and anus.   Now all animals show evidence of
having passed through some such stage as this gastrula, and the interpretation of the fact can mean but
one thing.   If embryology repeats past history this
must mean that near the beginning of the development of the animal kingdom, there was a common
ancestor which corresponded in its fundamental features with this gastrula stage, and further, that all
animals which at the present time show traces of this
stage in their development have descended from that
common ancestor.   To the common ancestral stage
thus indicated, Haeckel gave the name Gastræa, a
name which is seen to correspond to the embryological stage, the gastrula.   This name we will retain in
the following discussion, although it is very probable
that the Gastræa, as Haeckel understood it, never
existed.   The idea as Haeckel conceived it has

undergone some considerable modification within recent years, but these modifications do not alter the wide significance of the common appearance of this stage in the embryology of animals. While then the Gastræa as originally conceived may be modified in future years, there is no probability that future study will make any less significant the early ancestor whose fundamental structure was essentially that of the gastrula.

We may start the history of multicellular animals then by supposing that the unicellular animals first became aggregates of cells, and then that those cells which were situated on the outside of the mass acquired special functions connected with the relation of the animal to the external world, while the cells which were in the interior of the mass took for their duty the digestion of food which was passed to them from the exterior. In other words, there was a division of labor between the cells of the animal, and when this had appeared, there arose the first true multicellular animal. While we have no fossil evidence of the form of the first multicellular animal which appeared in the world, the evidence derived from embryology tells us pretty surely that it must have been one which could be described as a two-layered sac, probably open at one end, for the injection of food and the ejection of waste, whose inner lining served for digestion, and whose outer covering possessed motor and nervous functions.

As a further confirmation of the conclusion thus reached from embryology, it is highly important to notice that there is still in existence, among the

lower marine and fresh-water animals, a type scarcely
more than such a two-layered sac. The Cœlenterata
is perhaps in some respects the lowest group of
multicellular animals, and among this group some
orders have a structure which is almost exactly that

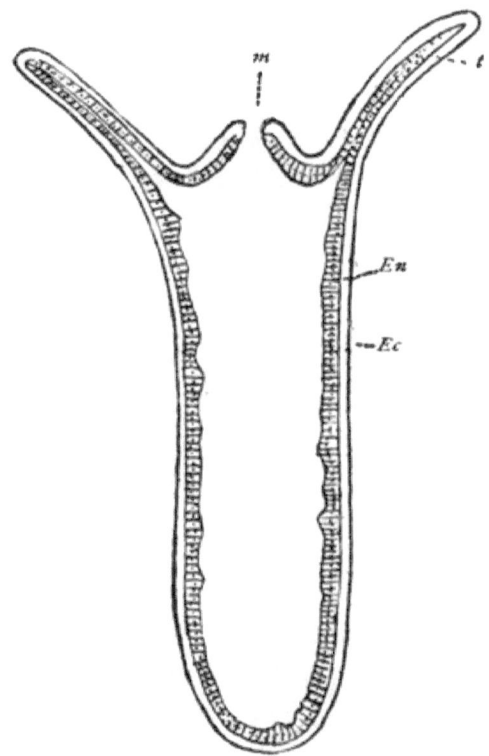

FIG. 7. Diagram of the structure of Hydra.—It differs from the gastrula of
Fig. 6 only in having the body around the mouth expanded into tentacles, *t*.

above described as characterizing the first multicellular animal. Hydra (Fig. 7) is readily seen to
be simply a two-layered sac, with an opening at one
end. The functions of the two layers of Hydra are

the same as those supposed to have been in the early primitive animal. Among other cœlenterates this simple type is more or less modified, but it can always be seen to have the same fundamental plan. Since, then, the lowest existing group of multicellular animals is essentially the same in structure as the Gastræa, to which embryology teaches us to refer all animals, we must conclude that the former existence of the Gastræa, as the starting-point of multicellular animals, is one of the best attested facts of biological science. It is a conclusion that can be gainsaid only by denying completely the value of embryological and anatomical evidence, and this would of course be to deny the cogency of biological science completely.

The first step in the history of life toward the development of the higher animals may then be briefly summarized as follows : Some original unicellular animals, in the course of their repeated divisions, failed to separate completely into single cells after dividing, but the cells thus produced remained attached to each other. At first the cells were all alike, and had similar functions, but after a time the cells on the outside and those on the inside of the mass began to perform different duties. Those on the inside could no longer use any powers of motion, and the possession of sensitive functions would be useless, since they had no direct relations with an external world to excite the sensations. It is a law of nature that any power which is not used begins to degenerate, and therefore the internal cells soon lost their motor and sensory functions.

Since they were protected from internal injury, they could, on the other hand, better perform certain functions connected with the preparing of the food, provided that they could succeed in getting hold of the food itself. The outer layer of cells, coming into direct contact with the world, retained all of the powers which enabled it to be stimulated by that world, and soon learned to pass food to the inner layer of cells for digestion.

We are still in the dark as to the exact manner in which the special form known as the Gastræa arose. According to some embryologists, the cells first formed a hollow sphere, and then one side of it was infolded, as one would push in the side of a hollow rubber ball. According to others, the mass of cells was solid, the outer ones soon becoming different from the inner ones, and later a cavity appeared in the middle, which broke through to the exterior at one end, and this opening formed the mouth. According to still another view, the cells at first formed a flat mass of two layers of cells, and this by folding up into a cup, or going through other modifications, became the two-layered sac which forms the first distinct stage in the development of the multicellular animals. According to others still, and this is the most recent view, a hollow sphere was formed, and thus cells from the shell migrated into the interior, one by one, to form the internal layer. The mouth opening was developed later. But whatever difference there may be in our ideas as to the details of this formation, no biologist questions the fact that very early in the history of the living world there was developed an animal with an outer and an inner layer of cells, and a mouth opening, and that this form has been the starting-point of most, if not all, of the subsequent multicellular animals. We shall call it the Gastræa, although it may not have been exactly the same sort of animal as that to which this name was originally applied.

Divergence of Types.

The next step in our history of animals was one of great importance. The Gastræa became moulded into types which foreshadowed the animal world of to-day. Of the modifications of the Gastræa by which it became changed into the various types of higher animals, embryology and anatomy alone give us evidence, and even here the evidence is clear only in a few cases. There seems to be conclusive proof that the Gastræa was the last stage which was shared in common by the different groups of animals, and possibly some groups branched off even earlier than this typical Gastræa stage. From this point certainly a divergence took place which soon resulted in the formation of a number of animal types which we now recognize as the sub-kingdoms of animals. The details of this divergence embryologists have not yet fully mastered. The simplest case seemed to be along the one line which gave rise to the cœlenterates. In this line of descent the original Gastræa attached itself by the end opposite the mouth, and by then undergoing various small changes in form, such as production of tentacles and elongation of the body, produced the group of animals of which

Fig. 8, *g*, is an example, the group of hydroids including corals, jelly-fishes and sea anemones. In nearly all other lines of descent from the Gastræa there was soon acquired a new fundamental feature. The mouth, which originally served for the entrance of food and the excretion of waste matter as well, was replaced by a pair of openings, one serving for each function (Fig. 8, *b*). Just how this took place is perhaps still a little problematical. The probability seems to be that the one opening of the Gastræa elongated into a slit and then closed in its middle. The two ends of this elongated slit remained open, however, one for the entrance of food, and the other for the exit of refuse. With the subsequent elongation of the body these two openings became widely separated from each other. After this further modifications of the body arose. Developing a shell on its back (Fig. 8, *c*), it started along a line which has produced the type which we have called mollusks (snails, oysters). Along one line of descent this shell assumed a spiral twist, giving rise to the snails (Fig. 8, *f*). In another the shell became divided into two pieces, one on either side, giving rise to the clams, oysters, etc. (*Lamellibranchiata*). In another line it became elongated and divided into segments, and this gave rise to the large type of animals known as segmented animals (Fig. 8, *d*), (segmented worms, *Crustacea, Insecta*). An elongation and segmentation and subsequent modification in many important particulars, the chief of which was the development of an internal skeleton, produced the type of *Vertebrata*. The details of these changes, however, need not delay us. They are technical in nature and could only be understood by one acquainted with comparative anatomy and em-

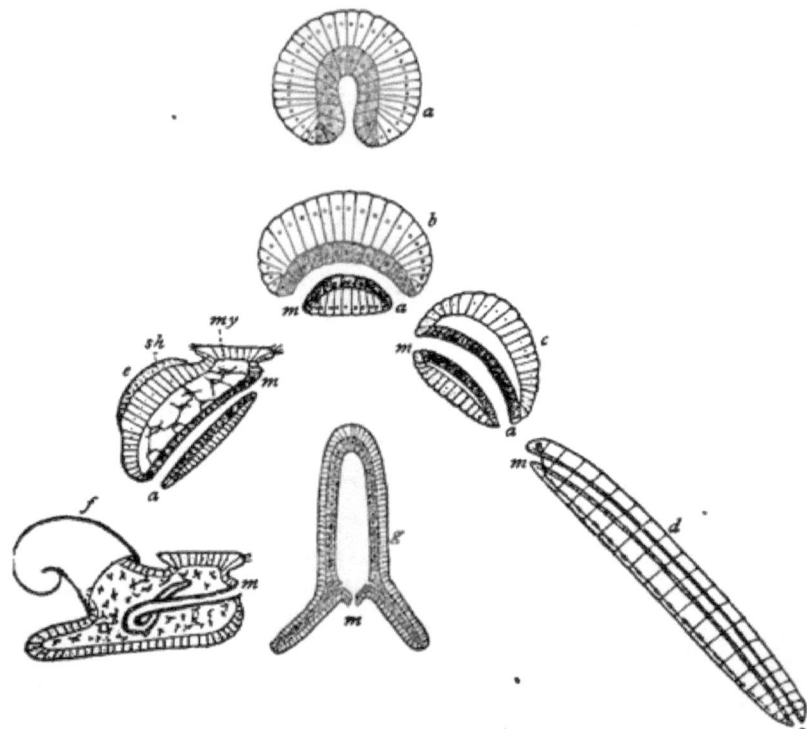

FIG. 8. Diagrams illustrating the origin of the Cœlentera Mollusca, and Annelida from the gastrula.—To produce the Cœlentera *a* becomes directly modified into *g*. To produce the Annelida the line of development is *a, b, c, d*. To produce the Mollusca it is *a, b, e, f*. In *f* the twist that appears in the shell twists the digestive canal. *m*, mouth; *a*, anus; *sh*, shell; *my*, cells which are to develop into muscles.

bryology. Nor, indeed, must it be understood that the embryologist can follow them in all cases even to his own satisfaction, nor that all embryologists agree. Various methods of accounting for the origin of the vertebrates are advanced. The relations of the star fishes *(Echinoderms)*, the group of low worms *(Vermes)*, and some other types still prove a puzzle to him, and it will doubtless require much investigation still before the embryological history is fully elucidated.

The history of the development of the primitive Gastræa into our modern types is a matter of speculative interest to the specialist, but to the general reader is too complex to make it worth while to dwell upon it. There is, however, concealed in this subject a general principle of development of the most extreme importance. From the history of this Gastræa we learn that the divergence of the great types of animals must have occurred early in the history of animals, and that *no new great types have appeared except in the early history*. The reason for this is easily seen. The differences which separate the great types of animals from each other are in points of fundamental structure and plan. Such differences could have appeared only in the descendants of some early form in which no special type had appeared. After the line of descendants had assumed a definite type, it is not likely that they would ever afterwards have changed their type though many changes in minor details may have occurred. Indeed, all evidence shows us that the types, after they have once become fully established, have not changed their plan but have remained constant. A tree when it starts from the ground as a seedling soon gives rise to several branches. Now this early branching which takes place in a few days after the seed springs from the ground, really determines the subsequent shape of the tree. No matter if the tree lives to be a hundred years old it will always be possible to see in its giant limbs the early branching of the seedling. These early branches become larger ones, and they in turn give rise to smaller branches and twigs, but after the first few weeks' growth it is impossible for the plant to produce any more primary branches. So in a modified way it seems true in the animal kingdom. Very early in the history of animals, the Gastræa trunk branched into several primary divisions, and these have continued to the present time. They have grown larger, they have produced many subdivisions, but the animal kingdom did not, after the early periods of its history, produce new primary divisions. In other words, no new sub-kingdoms have arisen since the earliest periods of the development. (For Vertebrata see Chapter IV.)

It is possible to reach this same conclusion from another standpoint. The development of the animal kingdom has been in all cases from undifferentiated to differentiated. Organs originally all alike and adapted to simple functions, have become different from each other and adapted to more complex functions. For instance, the mass of muscles which in the fish's body are adapted only to the flexion of the body from side to side, and the simple motions of its fins, become in the more developed vertebrates, like man, divided into several hundred separate muscles, each with its own function, all resulting in the complicated powers of motion shown by

6

his body. The development of animals has always been thus a differentiation. But we must bear in mind that only the undifferentiated can become differentiated, and it is plain that after an animal has once become differentiated in any direction the possibilities for further differentiation become immediately limited. Let us notice a single familiar example in illustration. At one time in the history of the horse family the possession of five toes on each foot was a common character. From a five-toed condition a large number of lines of descent were possible. But the actual animals entered upon a line of progress which resulted in the successive loss of four of their toes. Now with each step in this progression the possibilities for further development became limited. The animal with four toes was forever debarred from all lines of progression that required five, and the horse of to-day, having only one toe, has practically ended the possibilities of development in this direction. This one toe he must of course retain; and since it is a law of biology that an organ once lost is never redeveloped, any further differentiation of the toe of the horse is impossible. Now the same principle is true everywhere. As soon as the first step in any line of differentiation is taken, the possibilities for further development become immediately limited.

The most undifferentiated of all types of multicellular animals is the Gastræa. In this simple structure there are possibilities of an immense amount of modification, simply because it has developed no structure of its own. But just as soon as the descendants of this simple type become modified in any direction, the future development is immediately limited to the type thus produced. When some of the early animals modified their digestive organs so that there were two openings to the digestive tract instead of one, from that time their descendants were required to conform to this type. When some of them developed an elongated segmented body, it was no longer possible for them to return to the Gastræa stage, and start in a new direction. Advance in structure comes from assumption of different functions on the parts of organs originally alike, and as the parts of the body acquire a wider and more varied development, the possibility of further differentiation is more and more restricted. The further the development of animals progresses, therefore, the less will be the modifications of structure that take place, and the more must development be confined to the elaboration of details. This fact we find constantly illustrated throughout the history of animals, and it will be frequently referred to hereafter.

We can thus understand why the sub-kingdoms which arose early in the life of animals could continue to advance and elaborate each its own type by the production of many minor branches, but why none of them could give rise to new types which would present as great differences as the sub-kingdoms first appearing. If the Gastræa ancestor had remained in existence to serve in future ages as the starting-point of new lines of differentiation, then the production of great divisions would seemingly

have been indefinitely continued. But such does
not appear to have been the case. In other words,
the evidence of nature seems to indicate that the
different sub-kingdoms did not develop from each
other, but all (?) diverged early in the history of the
world from a simple undifferentiated animal which
bore some resemblance to the hypothetical Gastræa.
This conclusion is one of no little significance, for it
indicates that the origin of the large types of animals was a matter much more rapid than their subsequent elaboration. At the origin of the multicellular animals, a comparatively short time probably
produced divergence in the type which gave rise to
the great sub-kingdoms, while the millions of years
that succeeded only sufficed to elaborate and differ-
entiate these types. This understanding gives us an
explanation of the interesting fact that no new
types have appeared within the geological periods
since the Silurian, and it assists very much in understanding what seems to be the sudden appearance
of life in the oldest fossiliferous rocks.

Summary.

In the study of the early history of life three
points of importance stand prominently forth.

First: Far back at the beginning of life on the
globe, there was a period during which the unicellular organisms were the highest that existed. Already the plant life and the animal life had become
separate. We do not understand, however, that
this unicellular stage was the beginning of the history of life, for we are learning that there is a vast

field lying below the simple cell: but none of our sources of evidence give us as yet any sure traces of this earlier history. Of the unicellular stage of the history we are sure. Without being able to determine much in regard to them, we may perhaps look upon the unicellular organisms existing to-day as a tolerably correct picture of those of early times.

Second: We learn to picture to ourselves these early unicellular forms as multiplying by division, and we see among them many instances where the multiplication did not produce complete separation of the cells. The cells remained attached to each other and formed colonies of unicellular animals. Among these colonies a differentiation of the cells and of function began to make its appearance. To what extent different types of this early differentiation appeared, we do not know, but finally there arose one which has proved permanent. This was the two-layered cup to which has been given the name of Gastræa. This simple type with its mouth and stomach, proved itself strong enough to battle for life with nature, and it probably became the starting-point of all subsequent animals. The kingdom of plants separated itself permanently from the animals, and can no longer be traced with them.

Third; The Gastræa was the last point of common origin to which the different sub-kingdoms can be traced. From here there arose a divergence which soon produced the types of the existing animal world. Moreover, we have seen that not only is it impossible to trace the whole animal world to any later point of origin, but probably most, if not all, of the sub-kingdoms can be independently traced to this one point, indicating that this was the starting-point of all the animal kingdoms. It is doubtful whether even two of the great types existing to-day have had a common point of origin later than the Gastræa. This has been seen to be due to the fact that the great types are separated from each other by differences in fundamental plan of structure, and such fundamental differences could only have arisen from an undifferentiated form which had as yet developed no distinct type of its own.

CHAPTER IV.

THE RECORD FROM FOSSILS.

WE have thus far traced the history of animals through the unicellular stage to the Gastræa type, and we have seen that this Gastræa probably diverged rapidly, by various modifications in the shape and structure of its body, into different sub-kingdoms. Thus far our evidence has been only sufficient to indicate such general facts with no details. Embryology, with the assistance of comparative anatomy, has told us all that we know of this early history of living things, and these two sources of evidence, as already pointed out, can give no details. Still it is plain that the ground on which we are treading is much more sure than that which served as our footing, while trying to discover facts in regard to the primitive origin of life. While our knowledge as to the origin of life is all hypothesis, and we are not even sure of general facts, our knowledge of the divergence of the great types from the Gastræa is more than hypothesis, and of the general facts outlined in the last chapter we are tolerably sure. Their truth is commensurate with that of the embryological argument in general, and if we accept

the teachings of embryology as indicating the past
history of animals, we may regard the general facts
of the derivation of the Gastræa from the unicellular animals, and the subsequent modification of its
descendants into the different sub-kingdoms which
soon filled the world, as substantially proved.

We come now to the ages represented by fossiliferous rocks, and immediately the record is much
more definite. When we strike the fossiliferous
rocks we seem to be dealing with a more tangible
subject. The animals which were buried in the
ancient muds and sediments, and which we are now
unearthing, were actual animals and not hypothetical ones. With all of the cogency of argument from
the embryological standpoint, it cannot be denied
that the steps in the history of animals which it
points out are more or less hypothetical, and the
stages in the history of life of which it speaks
are stages of type only. But when we take a
fossil in our hands, we cannot question that the
fossil was once an actual animal, and an inhabitant
of the world in the earlier ages. It is now possible,
moreover, to determine many details in regard to
early life, for by means of fossils we deal with actual
animals, and not simply with structural types. For
reasons already pointed out, however, it is still
impossible to obtain connected history.

The Geological Ages.

Before proceeding with the history of animals as
taught us by fossils, it will be necessary to summarize
briefly the geological ages. The accompanying figure

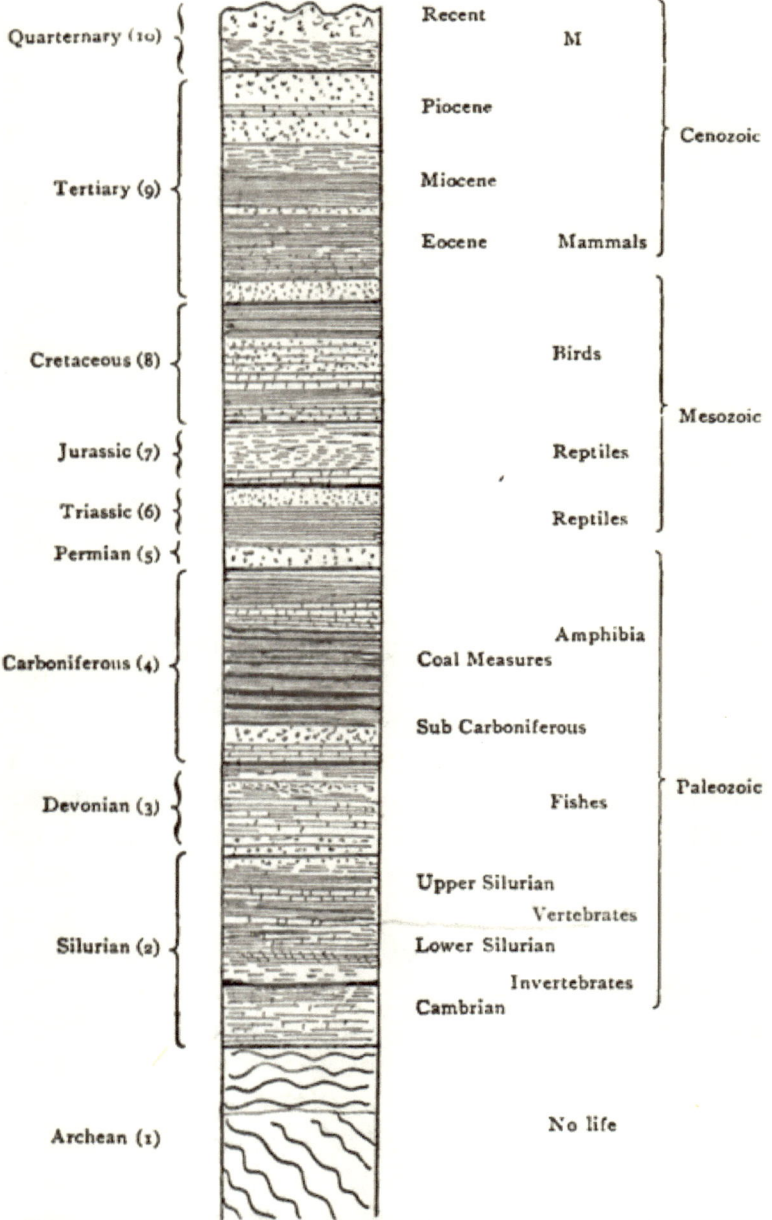

FIG. 9.—Diagram illustrating a section of the earth's crust to show the succession of the geological ages.

(Fig. 9) represents the ages in the order of their occurrence. In the subsequent pages the ages will be referred to by the names given in the left hand column in the figure, and for convenience of the reader who may not be familiar with them, a number will be placed after the name of the age which will indicate its relative order. Thus Archean (1) indicates that the Archean age was the first of the geological ages.

The following brief description of the life of the different ages will serve as an introduction to their more careful study :

PALEOZOIC.

The *Archean* (1) is characterized by having no traces of life. This does not mean that there was no life on the world at that time, but either that the conditions were not favorable for the preservation of fossils, or that the subsequent changes in the rocks (metamorphosis) have obliterated them. There is every certainty that life must have been in existence, but no fossil remains of animals assist us to this conclusion. All of the history dwelt upon in the last two chapters must have occurred during the Archean age, for, with its close, all of the early history had been past.

The *Silurian* (2) is characterized by the presence of animals in great profusion. The animals characteristic of the age were the invertebrates ; for while it is certain that vertebrates were in existence before its close, they were but slightly developed in comparison with the invertebrates, and did not appear at its beginning, at least so far as present evidence goes.

The *Devonian* (3) is characterized by the appearance of the vertebrates in large numbers. It is frequently called the age of fishes, from the abundance of these animals. No higher vertebrates are known to have been in existence.

The *Carboniferous* (4) was an age characterized by its abundant vegetation. It was during this age that most of the coal beds were deposited. The most important additions to animal life were the Amphibia, and probably the Reptilia. The Amphibia were large animals, and were especially abundant.

MESOZOIC.

The *Permian* (5) was an age of which only slight records are left. It was an age in which important changes in the animal kingdom took place. The true Reptilia became more abundant, and it is probable that at this period the Mammalia first separated from a reptilian stock.

The *Triassic* (6) was especially characterized by the development of numerous reptiles, some of great size.

The *Jurassic* (7) again found the reptiles the most prominent animals. New reptiles appeared, many of them being of immense size. At this time we find the first indication of the birds, suggested by many bird-like reptiles. The first real feathered animal also appeared, a bird with remarkable reptile-like characters. In this age the reptiles reached their highest development.

The *Cretaceous* (8) was characterized by a slight

diminution in the abundance of reptiles, though
they were still the predominant type. Birds more
closely allied to our modern birds began to appear.

CENOZOIC.

The *Tertiary* (9) began the modern era. Modern
birds were found in abundance, and the reptiles
lost their prestige. True mammals (in distinction
from the marsupials) made their appearance either
at this time or slightly earlier. During the Tertiary
they rapidly developed into the orders which fill the
world to-day, and almost immediately became the
predominant type.

The *Quarternary* (10), the age of man, brings us to
the present time.

These long periods were of immense duration, but
we have no means of determining even approximately
how long they lasted. Nor do they by any means
represent the whole of geological time. The geological history was doubtless a continuous one, and
if we had in our possession the whole of the history
we should hardly have been able to divide it into
ages. The periods outlined above seem very distinct
from each other because they are separated by lost
periods of which no record remains to us. Between
each of the periods above mentioned there is a break
in our record representing the periods during which
most important changes took place in the history
of life, but of which we can never obtain any knowledge. We do not even know whether the lost
periods were longer or shorter than those of which
we have record, but, judging from the amount of

change that took place in animal life, they must have been at least as long. Owing to these lost records it frequently happens that the opening of the different periods shows a surprising acquisition of the new forms of life. The sudden appearance of new types of life at the opening of the Tertiary (9), for example, is doubtless due to the fact that we do not possess the history of the immediately preceding period ; and so also the sudden appearance of life with the Silurian (2) is in the same way due to the absence of any record of pre-Silurian life.

Life at the Beginning of the Fossiliferous Record.

We will now turn our attention more closely to the development of life during these ages. In the first place it must be noticed again that it is no longer possible to trace the history of life along a single road. The Gastræa was the last point which the various types of animal life had in common. From this point many of the types diverged, and the complete study of the history of life would lead us to follow each of them. This would, however, involve a mass of detailed statistics which would be largely unintelligible to the general reader. Instead of following the history of life through its numerous details, we shall try therefore to take a perspective view of it ; noting the general truths and laws illustrated by the course of development.

Let us retrace our steps, and study more carefully the history of life during the geological ages. First we may ask, What was the condition of the animal kingdom at the time of our first fossil record of it?

Upon looking at the world during the Silurian (2) age, we are immediately struck with the surprisingly great diversity of its animal life. The divergence of types of animals, for which we have found evidence in the embryology, had already taken place. The Gastræa had given rise to a number of lines of descendants, all of which had more or less completely lost the original Gastræa characters, and had taken the characters of the animal types of to-day. Not only had the large types made their appearance, but there had already been time enough for most of them to diverge still further, and produce their characteristic classes.

The diversity of life found in the Silurian (2) age has greatly surprised geologists. This surprise is, however, disappearing, as we learn from embryology of the rapid divergence of types at the beginning of life, and as we remember the long blank Archean (1) age, during which life existed without leaving any distinct traces of itself. It must be remembered also that the Silurian (2) age itself was of very long duration, and many of the forms of animal life attributed to this period were not in existence at the beginning of the era, but were developed during its progress. The scantiness of the remains makes it, however, impossible to determine with any degree of probability which of the animal forms came in later. In our study of this early period of life's history, we will consider the age a unit, always understanding that it is a unit of immeasurable duration, and if we had been able to watch its progress, many of the forms which appear so suddenly would have been found to be slowly developing during its progress. Before, or during the Silurian age, all of the sub-kingdoms of animals came into existence. Cœlentera were present in the shape of sponges, corals, hydroids. Among Echinodermata were found crinoids, star-fishes, and some extinct forms of sea urchins. Various mud tracks tell us that marine worms lived in these early mud flats. Mollusks were abundant, some related to clams and oysters (lamellibranchs), and others related to our snails and cuttle-fishes. Brachiopoda existed in marvellous numbers, and the closely allied Bryozoa were plenty. Crustacea there were, much like our shrimps and lobsters, and a special type (trilobites) was highly characteristic of the age. Scorpions were present to represent the air-breathing animals, and the existence of their sting tells us that other air-breathing animals (probably insects of some kind) had also appeared. Of the sub-kingdom Vertebrata the indications are scanty, though there is no doubt that they were in existence by the close of the Silurian age, for in the upper rocks of that period unquestionable fossil evidence has been found, and recent discoveries have shown them in the lower Silurian rocks, thus carrying them almost to the bottom. The earliest vertebrates could not have had any hard parts to be preserved, true bones and even cartilage being of more recent origin, and this explains their scanty remains in the lowest rocks, although there can be no doubt that the vertebrates must have been more or less abundant even in the lower Silurian times.

If we examine this fauna a little more closely, its high state of diversity appears to be even more startling. Not only were all of the sub-kingdoms represented, but, omitting the vertebrates, all but two of the classes as well. Two thirds of the orders of invertebrates, and quite a large number of the families which are in existence to-day, were then developed. It may be said that all of the invertebrates had already developed their large trunks, and the subsequent ages have in most cases only sufficed to elaborate the minor branching of the trunks thus formed. The Archean (1) age had been sufficient for the divergence of the types of which embryological teachings have so plainly given evidence.

A few figures will illustrate the fact just mentioned, that the divergence of type before the Silurian (2) age produced greater modifications than those that have occurred since. Of the *sub-kingdoms* of animals, all were in existence during the Silurian. Of next smaller groups, the *classes*, all were in existence whose hard parts enabled them to be preserved, with the exception of one or two small classes, whose shell is but poorly adapted for preservation. Of the *orders*, of animals again, that have left any fossil records, thirty-four are found in the Silurian rocks, nineteen being in the Primordial, (*i. e.*, at the very bottom of the series), while only twenty-five orders have appeared in the more recent rocks. Since some of these twenty-five orders have only slight skeletons, their preservation must be regarded as accidental, and their absence from the Silurian (2) rocks does not prove that they did not then exist.

Nine of the later appearing orders were insects which fed upon flowers, and could only have appeared after flowering plants ; and in general we notice that the later appearing orders were largely land animals. Besides the orders included in the above, there are thirty-four which, from the soft structure of their body, have left no traces in the rocks at any period. Of these, it is certain that a number at least must have been in existence in the Silurian (2), since their close allies are known to exist there. Carrying our study into smaller groups, and including now *sub-orders* in our figures, we find the later ages coming out in more prominence. Of the one hundred and five orders and sub-orders that have left any evidence of themselves in the rocks, forty-four are Silurian and sixty-one have appeared in the later ages. Here too we find that of the sixty-one later sub-orders many are terrestrial, and at least twenty are insects. It is further noticed that the larger number of sub-orders which are of recent origin belong to the higher rather than the lower orders of animals. Taking *families* into consideration, a larger number were late in appearing, though a number of our modern families date from the Silurian (2).

Thus we see that the Archean age produced the sub-kingdoms, the classes, and most of the orders of animals, while the subsequent ages have only produced the smaller divisions, giving rise to a far less divergence in type, but a much larger profusion of minor branches. Now from all this it follows that the study of fossils is unable to help us to the knowledge of the early history of any of the sub-kingdoms

7

except the vertebrates, since practically all of them had developed before the fossil record began. It is true that in the subsequent ages most of the other sub-kingdoms have very much expanded in variety of forms, and have in general seized upon larger and larger fields in nature. But in most of them the real advance has been slight, and in many there has been actually no advance, but only production of new genera and species. In the vertebrates alone has all of the development taken place during the period of which we have fossil record, and here alone can we read a definite history of progression, since here alone can we trace anything like a continuous history.

With the opening of the Silurian (2), therefore, the animal kingdom bursts upon us in a comparatively high state of development. From this time on there is a constant widening of types, a constant succession and disappearance of old forms and the appearance of new ones.

As already seen, the Gastræa was the last point in the history of life shared in common by all multicellular animals. From this point there was a parting of the ways, and it is therefore no longer possible to follow the history of the animal kingdom as a whole. It will be necessary to take up the different branches and follow them separately. Of course, the further we trace them the greater and more complex will become the branching, but it is not our intention to trace this history in very great detail. It will be necessary, however, here to take into brief consideration the history of the various sub-kingdoms of animals as they are known to-day. Inasmuch as this will necessarily lead to considerable detail which will be uninteresting to the general reader, this part of our subject may be omitted without any break in continuity.

PROTOZOA.

The unicellular animals are all minute, and most of them are entirely soft. For these reasons they are not well adapted for preservation as fossils, and only one of the three classes has been preserved in the rocks to any great extent.

The *Foraminifera* have usually a shell of lime, and therefore have left traces of themselves in all ages. Even in the Archean (1) there is a curious body (*Eozoan Canadense*) believed by some to be the remains of Foraminifera, though it is usually regarded to-day as a mineral deposit. With the Silurian (2), however, true foraminifers appeared unquestionably in abundance, and every subsequent age shows traces of their presence. There are two geological periods in which their deposits are of special importance. The immense chalk beds of Cretaceous (8) are made very largely of shells of foraminifers. This chalk is almost exactly the same in its formation as the so-called *Globergerina ooze* that is being deposited now at the bottom of the ocean, so that it seems that chalk is still being formed in our modern seas. The second large deposit of foraminifers is the Numulitic limestone of the Tertiary, an immense bed of European rocks covering thousands of miles of territory. We must not conclude that in these two periods the Foraminifera were any more abundant than in others, but simply that the conditions were more favorable for their preservation. It is remarkably interesting that the species existing in the Tertiary (9) and many of those of the Cretaceous (8) chalk are identical with species found living to-day, and when we go still further back in history the amount of change is very slight. Indeed, Dr. Carpenter says there is no evidence of any fundamental modification or advance of the foraminiferous type from the Paleozoic period to the present time.

A second order of the Protozoa, the *Radiolaria*, possess a skeleton of silica which is tolerably well adapted for preservation. These have been traced back as far as the Silurian (2), though it is not until we reach the Mesozoic (5, 6, and 7) rocks that their remains become abundant.

The other Protozoa, having no hard parts, could not, of course, have been preserved as fossils. There can be no doubt now of their existence through all of the geological ages, since, as we have seen, the Foraminifera and Radiolaria, which are the highest of the Protozoa, have lived during all these periods, and the lower orders of any group always appear before the higher ones. In general, then, we may conclude with little chance of error that the Protozoa were in existence at the beginning of the Silurian (2), in practically the same condition that they are now, and that the long ages since have seen very little modification in their structure. They have been practically stationary, and their development was almost wholly pre-Silurian.

CŒLENTERA.

Four classes of animals are to-day included under the head of Cœlentera.

Porifera (sponges).—These are the lowest existing multicellular animals. As would be expected, therefore, they are very ancient in origin. They are found in the lowest Silurian (2) rocks and have been found in every age since that time. Here also the probability seems to be that the sponges of early times were much like those of to-day, so that there has been very little real advance in structure. The variety of type was however much smaller in early times than to-day.

Hydrozoa.—This class includes the hydroids (see Fig. 7) and the well-known jelly-fishes. The jelly-fishes being so soft, it is, of course, impossible to say when they really appeared, their absence as fossils proving nothing. The first traces we have of them are in the Jurassic (7). Many of the hydroids, however, have a shell, either of lime or of some horny material, and they have been found in all ages. The Silurian (2) rocks contain them in abundance. But the hydroids of this period were quite unlike those found at the present day. One entire class found in the greatest abundance at that time has since entirely disappeared (*Graptolitoidea*), the last traces being found in the Silurian (2). Several smaller groups have also disappeared. Of our modern forms, only one class (Thecophora) was in existence, and this was quite different from its representatives to-day. This class continued with little change during the whole of the Paleozoic (2-4) age, and it was not until the later part or the Mesozoic (5-8) that the modern forms began to appear, and the class assumed its present condition. How far this late origin of most of our existing groups is due to the lack of preservation of animals which really existed in the early ages, cannot be positively determined. The great probability is that most of our modern orders are much older than their fossils would lead us to believe.

Actinozoa (this includes the corals and the sea anemones).—The corals have commonly a lime skeleton, and are quite well adapted for preservation. They have been very abundant in all ages. The Silurian (2) rocks contain five of our modern orders, though most of the corals were quite unlike their modern representatives. The orders most abundantly represented at that time are, however, not the orders most abundant to-day. During all the Paleozoic (2-4) they continued to live in abundance, a large variety of forms being in existence all through that time. With the Mesozoic (5-8) however, we find the modern forms of corals becoming more abundant, and from that time the production of the modern coral was a matter of slow and constant growth.

The Ctenophora are wholly unrepresented as fossils, owing to their soft jelly-like composition.

In general, then, the Cœlentera form an ancient group, appearing in abundance in the earliest rocks, and being numerous in all ages. During the Paleozoic, however, the types represented were not those most abundant to-day, and many of them were confined to the Paleozoic (2-4). The modern Cœlentera seemed, so far as our record goes to-day, to have appeared with the Mesozoic (5-8), some of the ancient types disappearing entirely at that time and others becoming of subordinate importance. At this time also there was a great multiplication of sub-orders and families.

ECHINODERMATA.

This sub-kingdom is divided into five classes: *Crinoidea, Asteroidea* (star-fishes), *Ophuroidea* (brittle stars), *Echinoidea* (sea urchins), and *Holothuroidea.*

Crinoidea (stalked echinoderms).—This class may be especially regarded as the fossil class of the group, for although some are still in existence to-day, they are very few in numbers (only eight genera), and mostly confined to deep seas. In early geological times they

were, on the other hand, by far the most abundant of the echinoderms. With the beginning of the Silurian (2) great numbers and
varieties were found in existence. The class was then divided into
three orders (*Brachiata*, *Blastoidea*, *Cistoidea*), all of which reached a
high degree of expansion in the Paleozoic (2–4). After thus reaching
a culmination they gradually become less numerous and continued
to diminish until to-day, when there exists only the small number

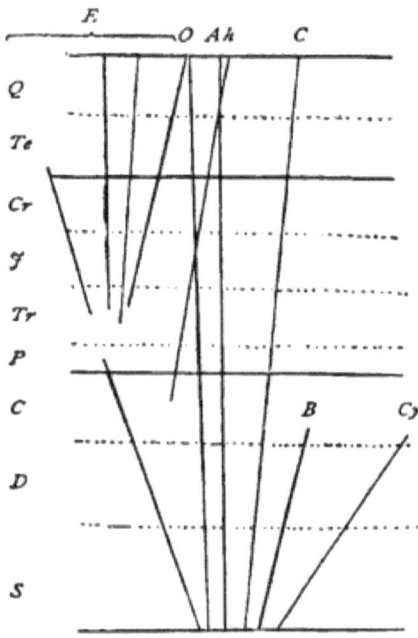

FIG. 10. Diagram illustrating the history of the Echinodermata.—The letters
on the left indicate the different geological ages. *E* Echinoidea. *O* Ophiuroidea.
A Asteroidea. *H* Holothuroidea. *C* Crinoidea. *B* Blastoidea. *Cy* Cystoidea.

above mentioned. Two of the former orders disappeared completely, and thus these that are left are simply surviving fragments of
a once predominant type. The modern crinoids, which are in some
respects different from the older ones, made their appearance in the
Triassic (6).

Asteroidea (star-fishes).—This class was also in existence in Silurian (2) times. The type of star-fish found in these rocks is slightly

different from that of the modern animals. The latter appeared, however, in the Devonian (3), and the ancient type disappeared with the Paleozoic (2–4). The modern forms of star-fish appearing in the Devonian (3) continued with little modification until the Jurassic (7), when the more strictly modern families began to appear.

Orphiuroidea (brittle stars).—The brittle stars were quite abundant in the Silurian (2), some of the genera being identical with those living to-day. Their history has been of little special interest. They have simply continued to expand slowly into the present condition.

The *Echinoidea* (sea urchins).—The sea urchins were well developed in the Silurian (2) rocks, but were of a type quite distinct from the modern forms (having more or less than twenty rows of plates). This order of *Palœchinoidea* continued to exist during the Paleozoic (2–4), practically disappearing with its close. With the Mesozoic (5–8) the modern urchins (with just twenty rows of plates) appeared, quickly expanding, and reaching a profuse state of development in the Jurassic (7) and Cretaceous (8). Since then they have been on the wane.

Holothuroidea (sea cucumbers).—The sea cucumbers have either no shells, or sometimes a few calcareous plates. As fossils they are known only by specimens of these plates which are occasionally found. No traces of them are found earlier than the Carboniferous (4), though the difficulty of determining their presence makes it not improbable that they existed at an earlier period.

In general, then, the echinoderms were very early developed. All of the classes, with the possible exception of the holothurians, were found in the Silurian (2), and were as radically distinct from each other then as they are now. The Paleozoic forms were on the whole, however, quite distinct from the modern representatives. The modern types appeared during the Mesozoic, and the group is at present on the wane.

MOLLUSCOIDEA.

Under this head are included to-day two groups of animals, whose position in the animal kingdom is unsettled, though they are certainly related to each other.

Brachiopoda.—These include animals externally resembling bivalve mollusks. To-day they are few in numbers and do not form a very important group, but in earlier times they were very abundant.

In the Silurian (2), indeed, the *Brachiopoda* were more abundant than any other group of animals, and the age is therefore sometimes called the age of *Brachiopoda*. At this time they reached their culmination, and have been declining ever since, though quite a number of them are still in existence. It is especially interesting that some of the genera existing to-day are not to be distinguished from those of the very earliest rocks. (*Lingula, Terebratula.*)

Bryozoa or *Polyzoa*.—These form a group of small animals not generally familiar except to students of natural history. They are abundant to-day at the sea shore, and usually are mistaken for plants by those unacquainted with them. They are small animals, but many of them have a calcareous shell. They appeared in the Silurian (2), and existed with no special change, though as the modern era approached, the animals gradually assumed more the type of the existing *Bryozoa*. Like the *Brachiopoda*, they form a fossil order whose chief development occurred in the past, although they have not yet become so widely extinct.

Mollusca.

Of all animals, we have the most complete geological record of the mollusks. They were well developed as a group at the beginning of the Silurian (2), and therefore their fossil record cannot tell us anything of their early history, but of their history since the Silurian we have a very full account. Their hard shells and aquatic habits have adapted them especially well to preservation. We recognize five classes of mollusks, whose history has been as follows:

Lamellibranchia (oysters, clams, etc.).—This group is characterized by having two shells, and is common both in fresh and salt water. The class was abundantly represented in the earliest Silurian (2), many of our modern families being already in existence. Indeed, quite a number of the genera then living still exist. By the close of the Paleozoic (2–4) are found many others not so well known. With the Mesozoic (5–8) there was a change. Many of the old forms disappeared and new ones took their places. During the rest of the geological ages there was an approach to the modern fauna. The old forms did not entirely disappear, however, and some of the oldest families continue to exist to-day. The Mesozoic is marked by the development of the clams, quohogs, and other less well-known mollusks. The modern forms have especially developed what is known as a siphon.

In general the Paleozoic (2–4) lamellibranchs were commonly without siphons, although some of the siphonate type even then existed ; the modern forms, on the other hand, are siphonate as a rule, though many of the asiphonate forms still exist. The most remarkable point in the history of this class was the development of a very peculiar order in the Cretaceous (9). This order *(Rugosites)* is very

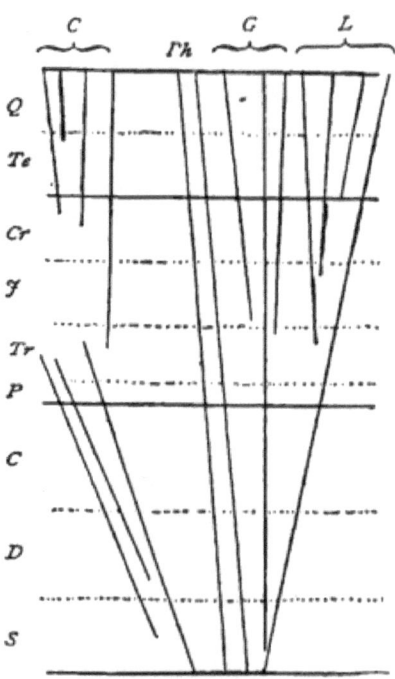

Fig. 11.   Diagram illustrating the history of the Mollusca.—*C* Cephalopoda. *P* Pteropoda. *H* Heteropoda. *G* Gasteropoda. *L* Lamellibranchiata.

unlike any other mollusks, and it suddenly appeared without warning in the Cretaceous and disappeared at its close.

Gasteropoda (snails).—The history of this class is quite parallel to that of the last. Beginning with the Silurian (2), they have been at all times very abundant. With the Mesozoic (5–8), the older forms began to disappear and the modern type became abundant. The first water and land snails are found in the Carboniferous (4).

Pteropoda (winged mollusks).—This is a little-known class, found to-day chiefly in the high seas. Pteropods were abundant in the Silurian (2) rocks, the genera seeming to be identical with those of to-day. The group has remained practically unaltered till to-day, except that some of the early families have entirely disappeared. At no time has the class been of much importance in the world.

Scaphopoda (tooth shells).—This is also a small and unimportant class, consisting to-day of only three genera. It is of great antiquity, however. The first remains are found in the Devonian (3), although the class probably existed long before that era. It has remained practically stationary.

Cephalopoda (squids, cuttle fishes, etc.).—This is by far the highest of the mollusks, and it has at all ages been an important group. The class has two well marked orders, one having four gills *(Tetra-branchiata)*, and the other two gills *(Dibranchiata)*. Beginning with the early Silurian (2), the Tetrabranchiata were quite abundant. Fourteen orders were then in existence. They gradually increased in abundance, size, and complexity, although only one new type was introduced during the Devonian, and no others until the Tertiary (9). They reached a very high state of development in the Mesozoic (8). From that time they rapidly diminished, and to-day only a single species *(Nautilus)* is left as a remnant of this once predominant type. *Nautilus* is in itself interesting as an example of a long-persistent species, being abundant as far back as the Silurian (2). Just before the tetrabranchs reached their culmination the first representative of the dibranchs appeared *(Triassic)*. These have continued to increase until the present time, one large section of them, however, disappearing with the Cretaceous *(Belemites)*. To-day the class, though existing in great numbers, is impoverished as to variety, only a few types remaining.

The mollusks in general will thus be seen to be a very old group, just about as well differentiated at the beginning of the Silurian (2) as they are to-day. The long ages have seen them always in abundance, and have witnessed slow changes and slight progression. There has been a constant expansion of the group, but it has consisted in the multiplication of families and genera. As elsewhere, the beginning of the Mesozoic (5–8) saw the older types giving way before the modern ones.

VERMES.

Of the history of the unsegmented worms we know nothing through geology, for their soft bodies have never been preserved. From their anatomical position we conclude that they are very ancient animals, and have doubtless lived through all geological ages.

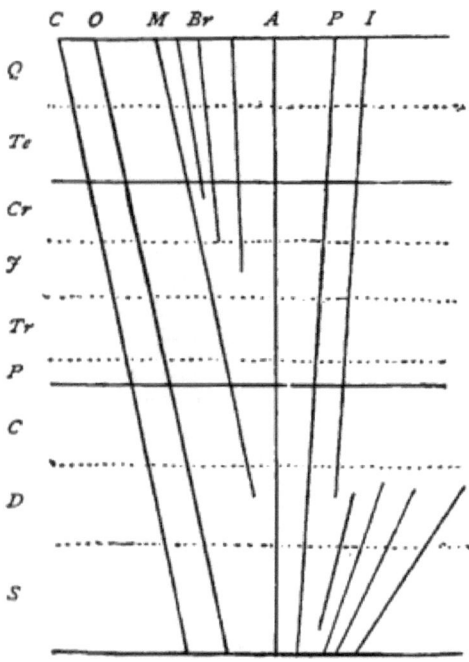

FIG. 12. Diagram illustrating the history of the Crustacea.—*C* Cirrepedia. *O* Ostracoda. *M* Macroura. *Br* Bracyura. *A* Amphipoda. *P* Phylocarida. *I* Isopoda.

ARTICULATA.

This large division of animals includes two provinces.

Anarthropoda (segmented worms).—The worms have soft bodies, not adapted for preservation as fossils. Some of them, however, live in tubes of lime or sand, and these cases are often preserved. Occasionally we learn of their existence by their tracks on the ancient mud. From such sources we know that they were in existence, well differentiated, during the Silurian(2) age, and have lived continuously until now, probably with little change.

Arthropoda.—This province again consists of three classes.

1. *Crustacea* (crabs, lobsters, etc.).—With the *Crustacea* we meet for the first time a class of animals, a considerable portion of whose development has occurred since the Silurian (2).   Still the class is an old one, and most of the lower orders were distinct at our earliest record of fossils.   The *Cirrepedia* (barnacles), *Ostracoda* (water fleas), *Amphipoda* (sand fleas), *Trilobita*, and *Euripteridæ*, as well as one order *(Phyllocarida)* which seems to have been the ancestor of the higher Crustacea, were all in existence during the Silurian (2).   But the trilobites and euripterids attained their culmination at that time, and soon after disappeared in the Carboniferous (4).   The higher orders certainly appeared later.   The *Isopoda* (sow bugs) and *Phyllopoda* first appeared in the Devonian (3).   The shrimps and lobsters are found in the Devonian (3) and Carboniferous (4), the crabs probably are in the Carboniferous.   Now since the crabs are undoubtedly derived from the lobster group, and this group from the Phyllocarida above mentioned, we have in these orders an instance where we can trace by fossils the origin of the modern orders from the earliest lower types. The Crustacea have always been abundant, and, with the exception of a few orders that have disappeared, are as abundant to-day as ever before.   The higher orders indeed were never so diversified as at the present.

2. *Arachnoida* (spiders and scorpions).—Under this head are found the oldest air-breathing animals.   The scorpions were in existence in the later Silurian (2), and true spiders are found in the Carboniferous (4).   That these have been in existence during all the later ages is therefore certain, though only scanty records have been found.

Myriopoda (thousand legged worms).—Being mostly land animals, these have left scanty remains.   They are found in the Devonian (3), and doubtless appeared earlier.

Insecta.—Though the insects form about five sixths of the animals of the world to-day, their habits prevent their ready preservation as fossils, and our history of the group is quite meagre.   Some of the lowest orders (cockroaches) were undoubtedly in existence in the Silurian (2).   During the whole Paleozoic (2-4), insects were abundant. All of them had a greater or less resemblance to the cockroach, although in the Devonian (3) and the Carboniferous (4), they took on features that allied them to the higher orders of insects, the beetles.

true bugs. and dragon-flies being thus foreshadowed. With the Mesozoic (5–8) the higher orders appeared, and by the end of the Jurassic (7), the modern orders of insects were all in existence. Of the various orders the higher ones appeared last. These orders, since they feed upon flowers, could not be expected until flowers themselves appeared, and this was not until the middle of the Mesozoic (5–8). [See Chap. VI.].

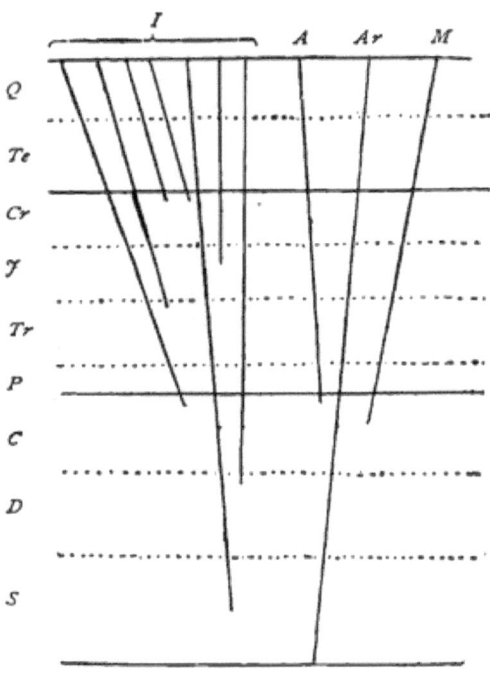

Fig. 13. Diagram of air-breathing Arthropoda.—*I* Insecta. *A* Araneina. *Ar* Arachnoida. *M* Myriopoda.

The insects have thus developed within the period of fossil history. In the Silurian they were represented by the lowest order and this order, during the rest of the Paleozoic, gradually diverged in the several directions which gave rise in the beginning of the Mesozoic (5–8) to the beetles and bugs and dragon-flies. Later, when the flowering plants made their appearance, there was a second divergence and a production of new forms depending upon flowers for their existence.

It seems probable that the remarkable social habit of the ants and bees did not develop until later than the Cretaceous (8).

The earliest traces of the Vertebrata are in the rocks of the lower Silurian (2). At that time there is no doubt that some forms of fishes were in existence, though the remains are rather scanty. The vertebrates then existing were low fishes whose skeleton was poorly adapted to preservation. Indeed, in the whole Silurian the traces of vertebrates are rare, although there can be no doubt that they were in existence. With the Devonian (3), they became very numerous. The Devonian seas were filled with large numbers of fishes, chiefly ganoids and elasmobranchs (related to the gar-pikes, and sharks). During this period these orders of fishes became very abundant and highly diversified. Late in the Devonian age some of the ganoids would seem to have acquired the habit of living partly in the air, and the habit thus acquired gave rise to the amphibians, which appeared for the first time in the next age, the Carboniferous (4). During this age the amphibians became abundant, and towards its close they seem to have become more distinctly aërial and to have ceased to be able to live in the water. This produced the true land vertebrates. In the latter part of the Carboniferous (4) or the Permian (5), we find that true reptiles had come into existence. Having once assumed a terrestrial habit, the reptiles rapidly expanded to appropriate the large field open to them, and in the next two ages (Triassic and Jurassic) they became more and more abundant, and grew to immense size. While these first land vertebrates were thus expanding, one side branch of them gradually acquired wings and developed into birds, which seem to have appeared first in the Jurassic (7) and Cretaceous (8).

From the Carboniferous (4) the reptiles had been constantly expanding in diversity, in number, and in size. But with the Cretaceous (8) they were slowly, yet surely, giving way to another and better adapted type of land vertebrates. Way back before the Jurassic (7) period, probably in the Permian (5), one of the early generalized types of reptiles seem to have sent off a branch of descendants which produced their young alive instead of laying eggs, and nourished them for a longer or shorter time by secretions from the dermal glands of the mother. These animals were at first small and probably weak. But they continued to exist during the age of reptiles, as a comparatively unim-

portant group. It would seem that somewhere toward the close of
the Cretaceous (8), or perhaps earlier, these animals made a further
improvement upon their method of producing young. They devel-

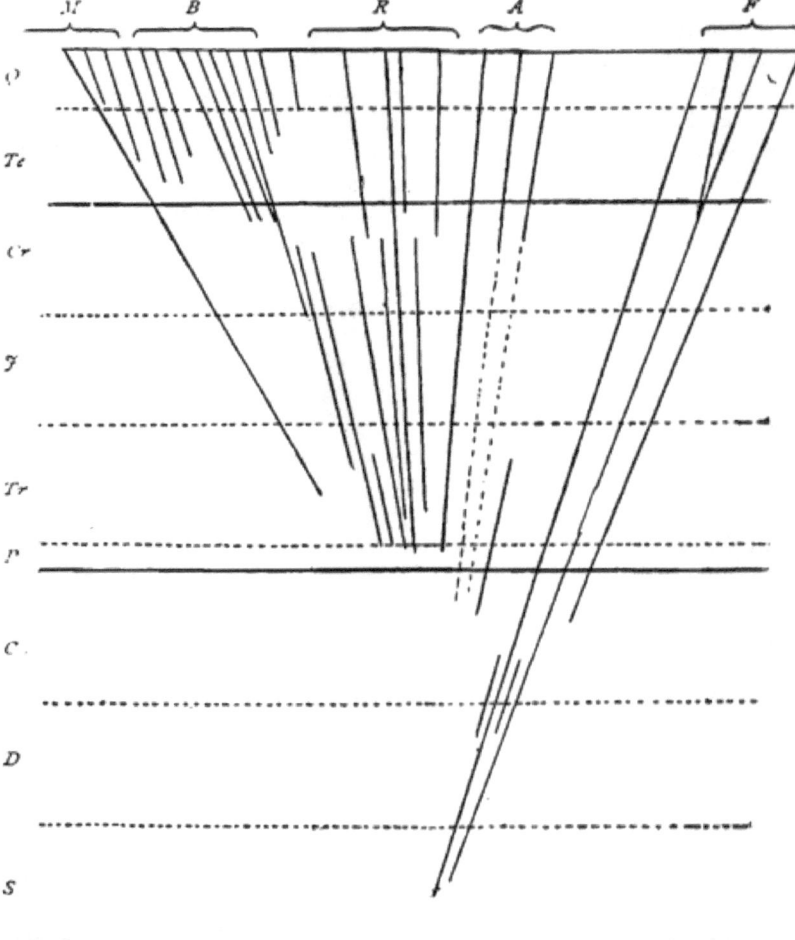

Fig. 14.—*M* Mammals. *B* Birds. *R* Reptiles. *A* Amphibiane. *F* Fishes.

oped the placenta which enabled them hereafter to carry their young
in the uterus till it was well developed. The placental mammals thus
arising proved to be a much stronger type than that which had existed

before.   Perhaps, too, the reptiles had expended the most of their energy, so that they were ready to retire before any newcomers.   At all events, we find that with the end of the Cretaceous (8) or the beginning of the Tertiary (9) the new group of mammals rapidly took the place of reptiles.   A large share of the reptiles disappeared and those that remained took a subordinate position compared to that of the new monarchs of the world.   With extreme rapidity now did the mammals develop.   In no other instance in the animal kingdom has development been so rapid.   Expanding into many types, occupying constantly new fields of nature, they soon began to assume forms familiar to us, and our modern families began to appear.   The mammal fauna thus became more and more like that existing to-day, until by the end of the Tertiary (9) the mammals of the present period may be said to have been in existence.   Finally in the Quarternary (10) period there appeared one animal who seized upon a new and untried field of nature for its own.   This new field was that of mind, and this new animal soon distanced every other competitor and became immeasurably superior to all other animals.   This of course was man.

CHAPTER V.

A VIEW IN PERSPECTIVE.

AFTER this outline sketch of the history of the animal kingdom, we may now try to take a perspective view of the whole, in order to get a better understanding of the true significance of the history. Details of science are of little interest or of little significance until they are collected into groups and formulated into laws ; and a bird's-eye view gives us a clearer idea of relations than an elaborate study of details.

Perhaps the most striking point which forces itself to our attention is one already noticed, viz.: the great extent to which the development of the animal kingdom had taken place even prior to the beginning of the fossil period. In the outline of the history of animals as sketched in the previous chapter, it has been seen that all the groups of animals except the insects and vertebrates had already reached a high state of development before the Silurian. In many of them there has been almost no change since that time. In others, the change has been chiefly in the addition of new families and genera without any special modification of type.

Not a few of them reached their culmination early in the Silurian (2), and soon after became extinct. In other words, the sub-kingdoms and classes of animals, and, in many cases, the orders and families, were as distinct from each other in the beginning of the Silurian (2) as they are to-day. The divergence of types had fully taken place prior to the beginning of our fossil record. Except in the one division of Vertebrata, the development of classes and orders must be learned from embryology and anatomy ; the study of fossils is entirely inadequate to help us to any connected history.

Paleozoic, Mesozoic, and Cenozoic Epochs.

A second equally striking fact is that in the long history that has intervened between the Silurian (2) and the present time there have been two specially prominent dates. The first was at the beginning of the Mesozoic (5–8). During the Paleozoic, as we have seen, various types of animals had existed in abundance ; but, though the three Paleozoic ages were of an immensely long duration, the modifications of type and the production of new families during its progress were comparatively slight. With the beginning of the Mesozoic, however, there was a remarkable expansion and development of almost all classes of animals. Within a comparatively short time a greater expansion into new forms took place than had occurred during all of the Silurian, Devonian, and Carboniferous taken together. A glance at the diagrams (Figs. 10–14) will show this expansion in a striking manner. The fact that there was such

a general development of new forms at this period
would indicate that some highly important change
in the condition of the surrounding living world
occurred to act as a stimulus. Although it is impossible to say positively what these new conditions
were, we may perhaps find a partial explanation in
the reduction of the amount of carbonic acid in the atmosphere. During the Carboniferous (4) era, vegetation had been rapidly at work drawing the CO_2 from
the air and storing it up in the form of coal. Roughly
speaking, we may say that all of the carbon now
stored in the coal beds was in the atmosphere in the
form of CO_2, previous to the Carboniferous age.
Plainly, the withdrawal of this CO_2 by the Carboniferous vegetation rendered the air much better
fitted for the terrestrial life of animals, and it is
interesting to find that during this period or just
after it the first air-breathing vertebrates came into
existence. Though this may not be a sufficient
explanation for the whole of the marvellous expansion of type in the early Mesozoic, and especially
of marine types, we are nevertheless probably correct
in regarding it as one of the most important factors
in this expansion. At all events, it is true that some
impulse acted upon the animal kingdom at about
the beginning of the Mesozoic which produced an
exceptionally rapid advance and a divergence of
form.

It is especially noticeable that this era, which
was so marked in the life history of animals, was
not an era of especial note in the vegetable kingdom. It was not until toward the Cretaceous (8)

age that the plant world received a similar impulse, causing it to expand into modern forms.

The second date of importance in the history of animals was at the beginning of the Tertiary (9). With this age we may say that the strictly modern order of nature began. From this time we begin to find in abundance the families and genera of animals which characterize the world to-day. Of our existing species of animals, the lower ones were the first to appear. Or stated in other terms, the lower species of animals are of longer duration than the higher ones. In the lowest rocks of the Tertiary (Eocene), we find representatives of existing reptiles living contemporaneously with a world of extinct mammals. None of our living mammals had then appeared, though modern reptiles were abundant. Of the mammals, too, the lower orders approached the modern condition first, the Insectivora and the Edentata considerably antedating the Primates. During the Tertiary, the approximation toward modern fauna was very rapid. Old species disappeared, and new ones appeared leading directly up to the present time. The new era inaugurated by the Tertiary was not, however, so marked as that coming in with the Mesozoic, for except in the class of mammals there was far less expansion of type than occurred in the early Mesozoic. Still the new era is sufficiently well marked. Of the causes which produced it we have no satisfactory knowledge.

A Constant Change of Species.

Looking over the whole history as seen by our collections of fossils to-day, we find that through all this long series of ages there has been a constant change of species. Few species of animals have succeeded in remaining in existence very long; or, it is probably better to say, few species of animals have succeeded in breeding true for a very long time. We are, of course, speaking geologically, and thus are reviewing time by ages, rather than by years. Sooner or later the descendants of any one species died or became so changed from their earlier form as to constitute a new species. Thus throughout the past new species have been continually appearing. It is indeed very seldom that any species continues to exist from one age to another, and thus, as a rule, the species of each age were distinct. As already noticed, there were long periods between the geological ages, during which no record of life has been preserved. These blank periods were undoubtedly periods of considerable disturbance in nature, and they lasted many thousands of years. The only record we have of the events that occurred during these times is in the modification which they produced upon the living world. The disturbances were usually sufficient to change animal forms so much as to produce an entirely new set of species. It is thus usually possible to determine closely the age of any fossil by its specific characters.

Nevertheless, we find that no universal rule can be given; for while most species change with the advent of new ages, some of the species of animal, were much more persistent. Some forms of life remained with little change through all the geological periods. This is true of Lingula, some foraminifers, and some mollusks. The species that are found now are, to be sure, different from those found in the Silurian (2), but the genera are identical, and the differences between the species are not great. In such cases the same species may exist from one age to another. These are, however, exceptional instances. Generally the particular species characteristic of each age were replaced in the subsequent age by a new set, and this same fact is also true of the genera. Families were longer lived, and many families living in the Silurian (2) age inhabit our seas to-day.

It is not common, then, for any species to succeed in outliving the long periods between the geological ages. Nevertheless, the different species of animals have had quite different lengths of life. Some are confined to a single stratum of a single age, while others extend through two or three geological systems (some Brachiopoda). It is found also to be a general rule that the lowest species are of the longest duration. For instance, certain species of Foraminifera (Saccaminina Carteri) appeared in the Silurian, and have continued to exist until to-day with little or no modification. So, too, with the Lamellibranchiata and Crustacea, the lowest forms appeared early and continued to live a long time, while the higher ones were more rapidly modified. Again, as a somewhat better illustration, we find that in the Tertiary (9), the reptiles had already assumed the forms which have continued to exist up to the present time. Not so, however, with the higher vertebrates, which have undergone great changes since that time. The same is true everywhere. Low

animals with simple structure are not so delicately
adapted to their environment as to require a change
for every little change in their surroundings. The
higher animals, however, are so complex and withal
so delicately adjusted to their environment that any
little change of conditions throws them out of har-
mony, and requires change in structure to meet the
new conditions. A steam engine requires much
more care and gets out of order more readily
than a water-wheel. So the types of low animals
have continued to exist, while the higher ones have
been rapidly modified.

It is to be noticed further that not only has this
change of species been a constant change, but that
it has also been a gradual one. There have probably been no abrupt breaks in the history. We find
that when one species disappears, it is either replaced by another closely related form, or the whole
line becomes extinct. In all cases where the rocks
have preserved for us anything like a continuous
history, it is seen that there are no abrupt transi-
tions from one type to a radically different one. In
the few cases where the record is exceptionally complete, it is possible to trace one species into another
by innumerable connecting links. Such evidence
is, however, rarely forthcoming, and new species
commonly seem to appear abruptly. In many instances, indeed, quite new types of life seem to come
suddenly into existence, but wherever this is the
case we find that there is usually a break in our
record of history. Now if we take into account the
long periods of unrecorded history that intervened

between the ages, we are of course prepared to find that each age is characterized at its outset by a new set of species, and in some cases by a distinctly new fauna. Remembering, then, that all breaks in the history occur where there are breaks in the record, and that there are no breaks in the history where the record is complete, we may unhesitatingly conclude that the real history of life has been one of continual slow change, without breaks, with each age passing imperceptibly into the next.

Animal Types the Same To-Day as in Earlier Times.

With all of this extinction of old forms, it is remarkable that since the Silurian (2) no great type of animals has disappeared. All of the sub-kingdoms in existence in the Silurian are still in existence to-day, and the same is equally true of the classes and nearly so of the orders. It is marvellously interesting and surprising to find that all of the fossils found in the rocks deposited so many millions of years ago can be readily placed with one or another of the divisions of animals that are in existence to-day. The fact gives us a forcible lesson that the animal kingdom is one, and that during all the history of the world it has been a unit. It tells us that the same laws in force to-day were in force millions of years ago in the world. It indicates that some bond has united the animals of the rocks with those of our seas and lands ; and this unity gives us one of the strongest arguments for the belief that heredity and evolution have been the laws presiding over the development of the animal kingdom.

A second fact of equal significance is that not only has no large type disappeared, but no new great type of animals has appeared since the Silurian (2) age. This has been sufficiently discussed, and only need to be mentioned here as correlated with the subject in the last paragraph.

During the geological ages species differed in their range as they do to-day. Some were confined to small localities (certain species of trilobites), while others had an almost world-wide range (Foraminifera). As would be expected, the species with a long range in time were usually the ones that had also a large geographical range (Saccuminina Carteri). To this, however, there are exceptions, for some species confined to a single system of rocks have an enormous range in geographical area (certain ammonites).

In looking over the whole geological period we find that after a species of animals has once disappeared it has never again reappeared, no single instance being known which would serve as an exception to this rule. The principle may be carried out still further. No type of animals is ever a second time developed. It is probably even correct to say that throughout the wide animal kingdom no organ that has fully disappeared is ever again redeveloped. Animals cannot develop a second time that which they have once completely lost ; when a change of circumstances requires an animal to perform again a function formerly performed by a lost member, the lost member is not redeveloped, but some other organ assumes a new function. For instance, the insects once possessed many legs but they have universally lost all except three pairs. Now the larvæ of butterflies (caterpillars) need a larger number of legs than this, but instead of redeveloping the lost legs they have simply had a fold of skin modified to serve the function of legs.

Early Forms Intermediate Between Existing Types.

The next point of significance which we notice in the examination of the history of animals is that, although the earlier animals all conform to types still in existence, it is frequently impossible to classify them satisfactorily. Among the invertebrates there is not very much difficulty in this respect, probably because of the fact that even our earliest rocks contained the types well differentiated from each other. When we study the early representatives of the higher types whose history is more or less completely represented in the early stages, it becomes difficult and indeed impossible to determine where to place them. The early insects of the Devonian (3) and Carboniferous (4) seem to be related to several of the present orders of insects, and it is impossible to determine whether they are to be called beetles, bugs, and dragon-flies, or are to be regarded as all cockroaches showing an approach to these other types of insects. Especially is this difficulty of classification true in the case of the mammals. Among the early representatives of this class of animals in the Tertiary (9) some are so truly intermediate in type that it is impossible to determine whether they were insectivorans or marsupials, while others ·stand midway between the carnivores and ungulates. In short, many of the forms found in these rocks are so intermediate in type between the orders which we find in the world to-day that they cannot be regarded as belonging to any of them, and at the same time they show so many characters in common with the mammals now existing that it is equally impossible to regard them as forming distinct orders. They were, in fact, intermediate types. It is commonly true that while the fossils require no new types created for their classification, the fossils representing the introduction of any group show such a complication and combination of relations that it becomes impossible to classify them satisfactorily into any of the modern orders. As we study fossils of the succeeding rocks, however, we find the similarity of the structure to the modern orders more and more close, so that the nearer we approach the present time the easier becomes the task of fossil classification. The interpretation of this is of course found in the fact that the various orders of animals have separated from common centres in accordance with the principle of descent, or otherwise. The nearer we come to those centres the greater is the similarity of the animals; it was only in later times that the different lines of descent became sufficiently separated from each other to be recognized as distinct orders.

Introduction, Development, and Decline of Types.

We must now notice more particularly the history of some representative groups, studying their method of introduction, expansion, and extinction. In general we may say that, at the beginning of their history, all groups of animals are few in numbers and of slight variety; then they expand until a culmination is reached, when they begin to diminish in numbers until they reach extinction, and are replaced by other groups. Such seems to be the general history of all groups. Taken separately, however, the groups

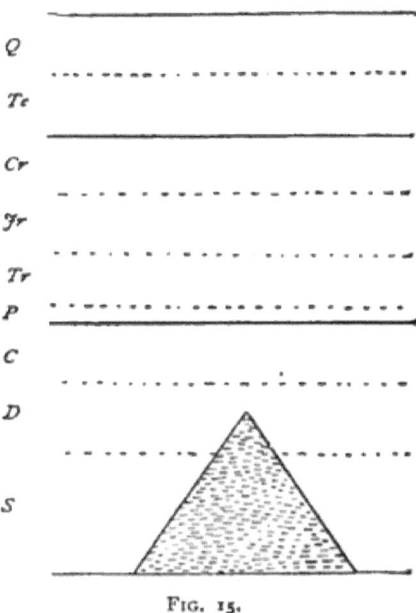

FIG. 15.

differ widely from each other, some of them having become extinct long ago, and others not yet seeming to have reached their culmination. Although the history of the different orders and classes is almost as varied as the groups themselves, still we may recognize several distinct types of development.

Fig. 15.—Some groups of animals were in existence in abundance at the beginning of our record; they

increased rapidly in numbers and diversity, soon
reached a culmination, and then disappeared with
an almost equal rapidity (see Fig. 15). The trilobites, for instance, were numerous in the Silurian
(2). During this age they became highly diversified,
reached their culmination in the last of the same
age, and then rapidly diminished in numbers and
disappeared. The same is true of euryptids. The
graptolites were abundant in the Lower Silurian (2),
reached their culmination during this age, and disappeared completely with it, no traces of them being
found in any subsequent period. Other examples
of the same kind are the Cystiphilloidea, disappearing in the Devonian (3), the Blastoidea and Cistoidea
in the Carboniferous (4), and also the Palæechinoidea
ending their history in the Carboniferous (4).

Fig. 16.—Some groups not in existence at the beginning of the history seem to have suddenly appeared
and then as suddenly disappeared again (see Fig. 16).
The rugosites of the Cretaceous (8) are the best examples of this. These remarkable mollusks, so unlike
any other representatives of the group, seem to have
become quite abundant in that age. They completed
their history, however, with the Cretaceous, and no
traces of them subsequently have been found. None
other of the invertebrate orders had a similarly short
life. One order of the dipnoids (Acanthodini), one
of Amphibia (Stegecephala), six orders of reptiles
(Ichthosauria, Plesiosauria, Pythonomorpha, Thero-
morpha, Dinosauria, and Pterosauria) have been
confined to comparatively short periods in the past.
Three orders of birds (Saururæ, Odontholcæ, Odon-
totormæ) and at least five orders of mammals

(Toxodontia, Condylarthra, Amblypoda, Mesotheria, Tillodontia) are confined to the Tertiary (9). The duration of the existence of the above-mentioned orders was short. In many cases the orders were confined to one system of rocks. Of course the sudden appearance of these various orders is, in a measure, misleading. They always appear after a

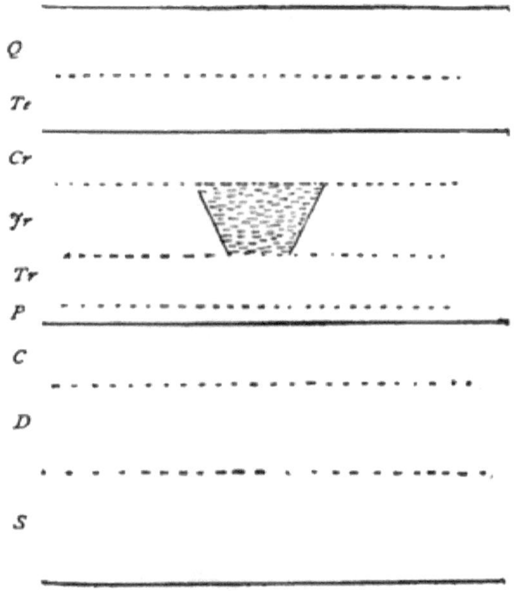

Fig. 16.

long period of unrecorded history, commonly at the beginning of a new system of rocks, and this fact must be interpreted as meaning that their seemingly sudden appearance is due to their earlier history having been lost. In some of the orders mentioned, indeed, the appearance was not very abrupt. But, nevertheless, all of these groups of animals are marked from others by their rapid development and sudden extinction.

Fig. 17.—Some groups have appeared in greater or less abundance in the Silurian (2), have rapidly culminated, and then slowly dwindled away, but have never become entirely extinct, even up to the present day (see Fig. 17). The crinoids form our best illustration. They were present in abundance early in the Silurian (2) rocks, they rapidly expanded, reached

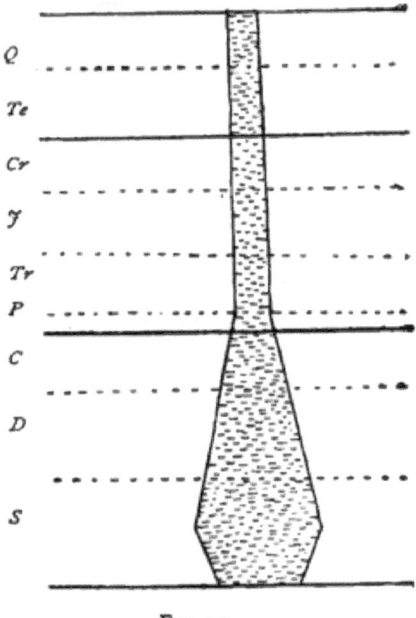

FIG. 17.

their culmination in the Carboniferous (4), and have since that time been constantly diminishing in numbers. They are not yet extinct, but only eight genera of this once predominant type are known to exist, and most of these are confined to the depths of the sea. The Brachiopoda also were so abundant in the Silurian (2) age that they may be called the characteristic animals of that era. Immediately after the

Silurian (2), however, they began to diminish in number, and have been growing of less importance ever since. To-day, though not so much diminished in numbers as the crinoids, they form a comparatively unimportant group with few genera. The orders of Phyllocarida form the only other important illustration. This is a small group of Crustacea, quite abundant in the Silurian, which, with the exception of a single genus (Nebalia), is to-day extinct.

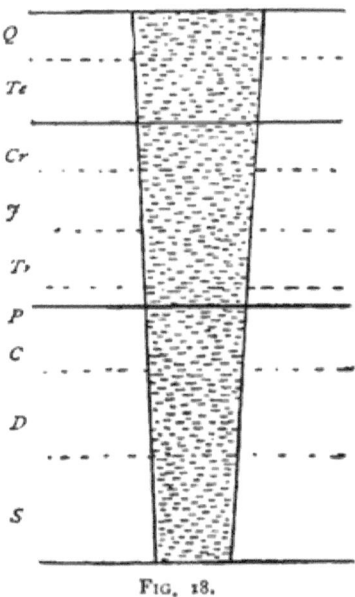

FIG. 18.

Fig. 18.—Some groups of animals appeared well developed with the Silurian (2) age, and have continued to exist in undiminished numbers ever since, or have even increased in number and diversity (see Fig. 18). This is true of a majority of the orders found in the Silurian. Many of them seem to have been constantly increasing in range with greater or less rapidity. This is true of the hydroids, corals, star-fishes, mollusks, and especially of the air-breathing insects.

Fig. 19.—Some groups not present at the beginning of the Silurian appeared subsequently, perhaps in small numbers, and from the time of their appearance

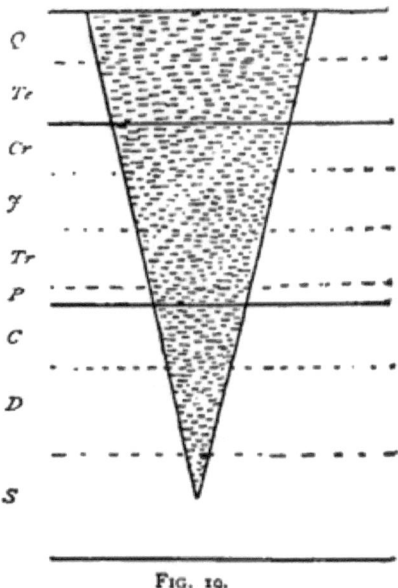

FIG. 19.

have continued to expand (see Fig. 19). Their numbers have grown larger and their diversity of structure has constantly increased. This has continued until to-day, and now they exist in greater abundance than ever before. Under this head are included all of the vertebrate orders living to-day, and nearly all of the insects, at least of the higher orders. Among the lower animals we find it true of the dibranchiate

9

Cephalopoda (squids), of an occasional order among
the rest of the mollusks, of our common lobsters,
shrimps, and crabs, of the order of modern sea-
urchins, and seemingly so of some of the orders of
Cœlentera (Alcyonaria), etc., though of this we cannot be sure, since the animals appear to be so poorly
adapted to preservation that their early history is
uncertain.

It is plain that all of the orders of animals existing
in great abundance to-day must come under one of
the two classes last mentioned, and it will thus follow
that nearly all of the animals well known to the general reader have had the history of constant expansion from the time of their appearance until to-day.

Fig. 20.—Not infrequently in the history of some
groups there has occurred a long period of stationary
condition, followed by a rapid development (see Fig.
20). A very striking illustration of this is shown in
Fig. 10, which represents the history of the echinoderms. From the figure it will be seen that the sea-
urchins appeared in the early Silurian (2), but during
this and the long Paleozoic ages remained practically
stationary. With the beginning of the Triassic (6),
however, some influence caused the urchins suddenly
to begin to develop into new forms. In a very short
time, comparatively, there were then produced a
large number of sub-orders and families, and the sea-
urchins were brought into existence practically as
we have them in the world to-day. By the time of
the Cretaceous (8) this development had ceased, and
from that time the echinoid group, as well as the rest
of the echinoderms, has remained stationary, or per-

haps has even been declining. The mammals again give a very pretty illustration of the same principle. This, the highest class of vertebrates, appeared first in the Triassic (6). These early mammals were all small animals and were of the lowest type, the marsupials. During this and the two subsequent ages,

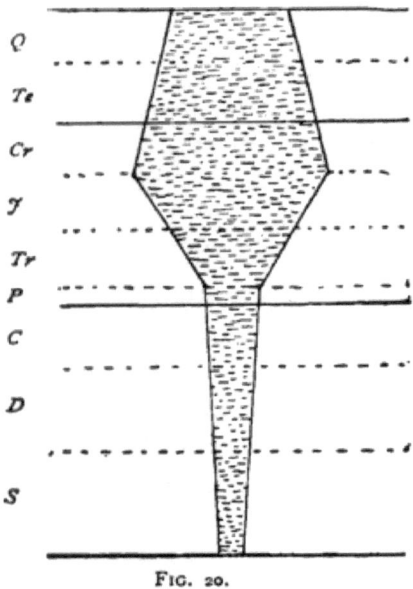

FIG. 20.

Jurassic (7,) and Cretaceous (8), the mammals probably continued to exist with very little change, the few remains found during these ages not certainly indicating any special advance over those of the Triassic (6). At the close of the Cretaceous (8), there seems to have been some influence acting upon the class which started them into a remarkably active development. Just when this impulse occurred, we do not know, but with the beginning of

the Tertiary (9) we find abundant fauna of the
higher true mammals, indicating that by this time
the mammals had been for some time under the
influence of an expanding force. From this early
Tertiary (9), the development and expansion of
the mammals occurred with great rapidity and in
a very short time, short at least compared with the
long period in which the class seems to have remained dormant, the higher mammals had diverged
into the modern ones.

Causes of the Sudden Expansions of Type.

The question, of course, arises why any group of
animals should have remained so long in a comparatively stationary condition and then have suddenly
expanded with such rapidity into a widely diversified
fauna.   It is hardly ever possible to give a definite
answer to this question. A somewhat general answer
can, however, be given, which is very suggestive
as indicating one of the important laws of animal
life.   It would seem that the comparatively sudden expansion of type has occurred when a group
of animals in some way begins to occupy a new
field of nature.   Where conditions remain constant,
animal life is held in comparative equilibrium, and
therefore is more or less stationary.   But any change
in conditions will disturb the equilibrium, and the
result is always a change in the structure of the
animals themselves. Now any change in climate,
in temperature, in atmospheric constitution, in configuration of the land and sea, etc., will be sure to
affect the equilibrium of life.   That such changes

have been constantly occurring in the past we have abundant proof, and it is certain they always induce changes in living nature. Indeed, the periods of greatest geological disturbances have always been those of most rapid evolution of animals (*e. g.*, the end of the Carboniferous (4) and Cretaceous (8). It would seem, however, that more restricted causes may frequently have been the origin of organic change. A group of animals may migrate into a new territory, and finding there a new field with a less powerful set of competitors, it will be able to expand itself to a much greater extent than in the old home. That this is a result of such migration is well known from the study of animals to-day. Undoubtedly such local changes have occurred in the past and have constituted one of the factors which have caused the periodic expansion of animals.

There is, however, another factor probably of even greater importance, but one which cannot be so easily understood or explained. Internal changes in the organism itself will produce new fields of expansion. It is well known that animals are constantly varying, now in one direction and now in another. These variations are continuous, and seem to have no direct relation to any change in environment. Of the many variations that appear, some are useless, and soon disappear. Others are of some value, and will be preserved by natural selection. Now it will occasionally happen that the variations may be of such a character as to make a radical change in the organism and fit it for entirely new conditions. For instance, we may suppose that all during the

Triassic (6), Jurassic (7), and Cretaceous (8) ages the
mammals were undergoing constant changes, and
were subject to numerous vicissitudes. Nothing
occurred, however, to give them any special advantage over their conditions or their enemies, and
they therefore continued for this long period with
little modification. But toward the end of the Cretaceous age it chanced that among the multitude of
variations some of the individuals acquired a new
character in regard to the habit of reproduction.
This new character gave them the power of retaining
their young in the uterus for a longer time than before,
carrying them indeed until they were well developed.
Now we know from the study of animals to-day that
this character, together with others correlated with it,
make the animal thus favored very much more powerful and far better adapted to contend in the struggle for life. We may not be able to say why it is,
but it is certainly true that the mammals with this
new habit in reproduction always triumph over the
marsupials. It is easy to suppose, therefore, that
as soon as this variation occurred in the mammalian
stock there was an immediate impulse given to the
development of the group. The mammals became
more active, and soon proved themselves more than
a match for the large reptiles. In the struggle for
food they soon triumphed over their marsupial pre-
decessors, and finding thus the whole world open to
their conquests, they multiplied rapidly and gave
rise at once to the multiplicity of types found in the
early Tertiary (9). A new field had thus been offered
by a chance (?) anatomical variation.

Now in the explanation thus given we must of course deal somewhat with hypothesis. We do not know positively that it was the occurrence of the new variation in the reproductive habits that gave the stimulus to the mammals, and thus caused the rapid divergence in the Tertiary period. We know, however, that the mammals did at that time receive a new impulse from some source, and did immediately expand to fill the new field of nature open to them, and we have pretty good evidence, moreover, that at the same time the above-mentioned change in the reproductive organs occurred. From all of this we learn to picture to ourselves the animal kingdom as constantly searching after new fields for expansion. The different types of animals are constantly migrating here and there, constantly subjected to new conditions, and therefore constantly undergoing change. Generally the severe competition with nature and with enemies keeps them in restraint by cutting off all sportive branches, and allowing only the central strongest forms to continue to exist. But occasionally a change in the conditions produced by geological forces gives certain classes the advantage over others, or changes in the configuration of the land produce new fields for expansion. Or, again, some one of the sportive branches proves to have inherent qualities of great vigor, and in this way offers a new field of organic type. In any of these cases the opportunity is seized upon immediately, and shows itself in rapid expansion. It is certainly a fact that the multiplication of species is greatest in the early life of a genus.

As already stated, it is very seldom that we can determine what were the circumstances which have produced the rapid development of certain types. We may suppose that the marvellous expansion of the reptiles in the Mesozoic was due to the fact that these were the first true air-breathing vertebrates, and had therefore the whole land to themselves. Such an unoccupied field would undoubtedly have had a tendency to stimulate them into rapid expansion. We may suppose also that the rapid appearance of the numerous order of insects in the Jurassic (7) and Cretaceous (8) was correlated with, if not caused by, the appearance of flowers and the acquiring of the habit, on the part of some insects, of feeding upon them. So too the rapid expansion of birds in the Cretaceous and subsequent age was due to the acquiring of aërial powers, which powers were of course due to internal variations in the direction of the production of wings. Perhaps the expansion of mammals was due to the change in reproductive habits as above mentioned. In a few other cases it is possible to make a guess as to the causes which produced expansion of certain groups of animals, but in general we must rest satisfied with ignorance of the details, and with only the general explanation that expansion is due to the occupancy of a new field in nature.

A General Advance from the Earliest Ages.

The next point for us to notice is that there has been a general advance in the animal kingdom from the earliest ages until now. Taken as a whole, there is no doubt that there has been an increase in complexity and diversity, and this is what is meant by advance. At the same time, when we attempt to follow this law into particulars it is by no means always possible to do so. In the sub-kingdom Vertebrata, there is no doubt of the facts. Beginning in the Silurian (2) with the lowest form, there followed in the Devonian (3) multitudes of fishes; in the Carboniferous (4) came the next higher class, the Amphibia, to be followed in the next era by the reptiles; these by the birds in the Cretaceous and the mammals in the Tertiary (9), and lastly by the appearance of man in the Quaternary (10). In all this we see a continuous advance. In the Articulata, too, it is possible to trace the same advance. Appearing at first in the form of low generalized trilobites and phyllocarides, the Crustacea developed the higher orders of lobsters, shrimps, and crabs considerably later. The insects seemed to appear in the Silurian (2) as cockroaches, which belong to the lowest order. In the other Paleozoic ages there appeared indications of the Neuroptera, and Coleoptera, and Hemiptera, which are always recognized as the lowest orders. It was only in the later periods that the higher orders made their appearance.

In the other groups of the animal kingdom, however, it is difficult to recognize any striking advance in structure. It is questionable whether we can say that the echinoderms of to-day are as a whole of a higher type than those of the Silurian, and a like question arises with respect to the Cœlentera, Mollusca, and Brachiopoda. At the same time, the

general increase in complexity even in the case of
these types is indicated by two facts at least. We
see that in early periods it was the lower orders that
were most abundant and widely diversified, while
to-day it is the higher forms that are predominant.
For instance, among the Articulata it was the low
trilobites and cockroaches that existed in abundance
and diversified profusion in early times, while to-day
it is the higher orders that are the most abundant,
even though in the insects the lower orders are still
in existence. So in the case of the echinoderms, the
most abundant order of early ages was the Crinoidea
although the other classes of echinoderms were all in
existence. To-day though the crinoids are still in
existence, it is the higher orders that are the most
abundant.

The second fact indicating the general advance is
the greater diversity of life to-day than in the earliest
times. Some groups have indeed passed into decline or disappeared, but most of those that do exist
are to-day in greater profusion than in any of the
past ages. Taken as a whole, it is one of the most
evident teachings of the history of life that there
has been through all the ages a constant increase
in the profusion of living things, and a continually
growing diversity of form. Even though it is possi-
ible to say that many of the families are really no
more highly developed than were their representatives in the past, still the fact of the increase of
diversity of type is a plain indication of a general
advance. Thus the fact of the evident advance in the
structure of the representatives of some types, the

increase in the profusion of the higher orders of all classes, and the general increase in the abundance and diversity of the families of animals which we can trace through all ages, shows conclusively that there has been a distinct advance in complexity of type from the earliest times up to to-day.

Although there has been thus a general advance, it is certainly true that it has not affected all orders of animals alike. We have already noticed that some have remained practically stationary in type. Some have actually gone down hill instead of up. Many groups of animals have advanced to a culmination and then begun to disappear ; and in the disappearing of animals we can frequently discern evidence of a reversal of the process of development, the later appearing animals becoming of a distinctly lower type than their relatives at the period of culmination. This has been specially well shown in the ammonites which passed out of existence in the Cretaceous (8). In many orders of animals the degeneration has been produced from other causes. Degeneration is always certain in the animal kingdom when any organ ceases to be used. Many orders of animals by becoming parasitic upon others cease to have any use for some of their organs. The result is a very general degeneration until the structure of the animal as a whole has become markedly degraded. Any of the orders of parasitic animals will serve as illustrations of this fact, and all of the sub-kingdoms show examples of such degradation. Among familiar examples may be mentioned fleas, which have lost their wings ; tape worms, which have lost all of their digestive organs, etc. Even the barnacles may be included under this head, for although not parasitic, they have acquired stationary habits, and have lost the locomotive organs they once possessed. While, then, the general history of the animal kingdom has been an advance, this advance is compatible with many cases of retrogression. All groups advance to a culmination, and then decline through decaying forms, and innumerable instances of parasitism have produced thus their inevitable degrading effects.

CHAPTER VI.

A VIEW IN PERSPECTIVE.—(CONTINUED.)

Soft-Bodied Animals.

LOOKING at the animal kingdom as a whole, we are now in position to trace something like a general outline of its history. Life appeared, very probably, before the land was fit for habitation, and consequently all animals were marine; at all events, it is certain that the earliest animals lived in the ocean. Now, drawing our conclusions from the study of embryology and comparative anatomy, we learn that the earliest animals were soft-bodied, and were provided with no hard parts for defence or protection. Their method of defence was probably the same as that of the low soft-bodied animals of to-day. The smaller forms were endowed with marvellous insensibility to injury, and with great powers of multiplication. Cutting them to pieces simply caused their multiplication, and this power of resisting injury served in the place of protective organs. The larger forms, like Cœlentera, were provided with stinging hairs for defence, and they also possessed remarkable powers of recovery from injury. Having no supporting organs, they were never large animals, but they doubtless filled the early seas in abundance.

Development of Hard Parts.

The next step in the history was the development of hard parts for support and protection. This for the first time made the existence of large animals a possibility. Almost immediately after the divergence of types of animals, the development of some form of skeleton began. The echinoderms developed plates of lime in their skin; the mollusks secreted upon their bodies a thick shell of the same material; some of the hydroids developed an external shell, and the corals a calcareous framework; the Articulata (except the so-called worms) developed upon the outside of their body a light but tough shell of chitin. Such skeletons as these, while they were certainly of great value, were in one respect a disadvantage. In the mollusks, for instance, they were large and clumsy compared with the size of the animal; they prevented rapid motion and effectually checked any great increase in size. It was only in later times, when the mollusks got rid of their clumsy shell, that any very great size could be attained, as in the case of the great giant squids. In the Articulata the shell, made of chitin instead of lime, was lighter and tougher, and therefore better adapted to activity. As a result, the Articulata succeeded in producing such active and well-protected groups as the insects. But the external shell of insects and crustaceans is not fitted for any large animal, the old trilobites (some of which were over two feet in length) reaching the extreme in this direction.

A little later another division of animals hit upon a new device for the support and protection of its soft parts. An internal skeleton was developed. This first appeared as a soft but resisting rod (noto-chord), inside of the back, and reaching from head to tail. This rod gave some strength, but could not become resisting without destroying the flexibility of the body. Therefore it soon became broken into segments to give the body flexibility, and then it hardened into a series of bones and formed the spinal column and the basis of an internal skeleton. Of this early history we have no trace from fossils, for it was not until this rod had hardened into cartilage that it was possible for vertebrates to be preserved in the rocks. The *early* history of the vertebrates, therefore, must be read from embryological evidence. This type of an internal skeleton immediately proved to be a success. This means of support gave strength and rigidity to the soft parts, and at the same time did not burden them so much as to render them unwieldy, for it was adapted on mathematical prin-ciples for furnishing the greatest strength with the least bulk.

The value of this skeleton immediately showed itself by the increase in size of the animals possess-ing it. We find that there followed now a new phase of animal history. For a long series of succeed-ing ages we see an increase in size among the higher members of the animal kingdom. Doubtless these early vertebrates had fierce contests with other ani-mals, but the large and active vertebrates soon proved themselves superior to the clumsy mollusks and the sluggish trilobites. The struggle of verte-brates with each other was severe. Size was a factor of the utmost importance in these contests, and through all the history of the vertebrates we find that the development of each class was marked by an increase in size. This tendency toward increase in size continued until the Jurassic (7), when it reached its culmination in the huge reptiles of this period, some of which were the largest animals that ever lived. Of course, we do not mean that every order of vertebrates became of large size, for many of the smaller ones did establish their right to live. Many small fishes and small reptiles did succeed in perpetuating themselves. Nevertheless we can recognize the truth of the general fact that from the Devonian (3) to the Cretaceous (8) or Tertiary (9) a tendency towards increase in size marked the history of the animal kingdom.

This increase in size, however, was sure to reach a limit. Great size is only to be possessed at the expense of activity and agility, and the great reptiles of the Jurassic probably became so large that it was a matter of greater and greater difficulty for them to procure sufficient food. As we have already seen, from the Tertiary period the mammals began to supersede the reptiles as the monarchs of creation. Even among the highest class of animals we find for a long time a tendency to produce animals of great size. Some of the mammal orders of the Tertiary (9) attained a size which almost made them rivals in bulk with the reptiles. But we soon notice in the history of this class a tendency to develop small size and agility instead of huge bulk. Of the edentates, the huge Megatherium disappeared, and only the small animals, such as armadillos and sloths succeeded in perpetuating themselves. The gigantic Dinotherium and the Mastodon were exterminated, and only two species of elephants remain to represent this once abundant type. In general, among mammals, it was the smaller animals, those which developed speed, as in the ungulates, or immense activity and strength in capturing prey, as in the Carnivora, that gradually became the most abundant, and thus eventually the predominant types. The development of bulk had been superseded by a new phase of progression. Increased activity, as shown by the power of flight, and the development of claws and teeth for defence, took the place of size as the most potent factor in the struggle for existence.

The Mental Factor.

But this phase of nature was soon to yield to another higher force. From the earliest record we have of animals we can see an indication of the final era in the history, *i. e.*, the era of mental activity. The brain of the lower vertebrates was small and indicated the possession of little intelligence. The same was true of the brain of early mammals. In some of them the brain was so small that it could be passed through the neural canal of the lumbar vertebræ, and was thus only a little larger than the spinal cord. From the early Tertiary (9) period, however, the size of the brain has been increasing. At the same time with its increase in size, the brain has been increasing also in complexity, the cerebral lobes becoming larger and showing more of a ten-

10

dency toward convolution. There is hardly a more significant fact in the history of animals, when we remember the approaching advent of man, than this increase in the size of the brain of mammals from the early Tertiary (9). Undoubtedly this increase in the size of the brain was connected with the increased activity of the mammals already mentioned, for activity and agility imply delicate control of the lower motor centres by the higher centres of the brain, and this requires more brain power. Active animals always have relatively large brains. But undoubtedly also the increase in the size of the brain was accompanied by an increasing amount of intelligence. In short, with the development of the mammals there had appeared a time when bulk had given place to muscular activity and agility, but at the same time the growth of the brain was gradually preparing the way for the time when intelligence should take the place of brute force.

Instinct.

Already, in an entirely different line of descent, do we find the mental nature becoming predominant. The higher order of insects has as its chief character the habit of living in colonies, and the consequent development of remarkable instincts. As a rule, the insects seem to have depended upon their powers of rapid multiplication as their chief means of defence. Most insects are weak animals, and all are small, but their immense powers of reproduction safely defend them from extermination. Some of the higher orders, however, learned to band

themselves together into colonies, and then developed a most complicated set of instincts, mostly
adapted for the preservation and integrity of the
colony. Now while instinct is indeed a mental
factor quite distinct from intelligence, it is, like intelligence, a function of the nervous system. The
insects thus first inaugurated the development of
the nervous system to an extent which made its
powers prominent factors in the preservation of the
race. At what geological date instinct became such
a prominent factor in the life of insects we cannot
say. The individuals found in the Cretaceous (8)
were all sexual individuals. Now the formation
of colonies is usually accompanied by the production of sexless individuals (neuters, workers). It
would therefore seem that the insects of the Cretaceous had not yet acquired their social habits, and
that the high development of instinct was subsequent
to this period. At all events it was a late event in
the history of insects, just as the development of
intelligence was a late event in the life of vertebrates, and it seems probable that instinct reached a
high development in the Tertiary (9) before the
special development of intelligence began.

Intelligence.

We are now prepared to recognize a period in
the history of animals when intelligence had become the predominant feature of nature. This of
course brings us to man, and we are now in position to consider the important question of the
relation of man to this history of life. For our pur-

pose there is no need of deciding whether man is a final step in the line of evolution or is to be regarded in some measure as a special creation. Man, by virtue of the powers he possesses, holds an important position in this life's history, and his position is the same whether his unique powers be regarded as evolved from those of animals or as newly created. The hint has already been dropped that the essential feature of man is the importance of his mental nature, and if we add to this his ethical nature, we have included all features of distinct importance that belong to him. Man may, indeed, be not incorrectly defined as the animal in which everything has become subordinate to the nervous system. For his existence in the world he depends not upon his physical force nor anatomical structure. He proves his right to live neither by defensive nor offensive armor, neither by superior strength nor superior powers of multiplication. The possession of mind alone is his pride. No sooner did his mental nature become a predominant feature than his physical nature ceased to undergo any considerable development. It is one of the curious facts of nature that man, who possesses powers of mind so infinitely superior to those of the highest animal that there is really no comparison between them, should at the same time be so much like the apes in his anatomical structure that naturalists think he ought to be classed in the same genus with them. Anatomically, then, man is an ape, while, so far as mental powers are concerned, he deserves a new kingdom to himself.

It is not difficult to find an explanation for this surprising fact. Among animals physical power is usually the only feature by which one animal conquers another. With them, therefore, when a new species is established it must be for some superiority in physical ability. As a result the anatomical structure of animals changes with every advance, for each new species must be in some respects better adapted to its conditions than the older one from which it came. With man, however, the physical side of nature is comparatively of no importance. From the moment that he proved himself master of animals by his intelligence, the development of his physical nature became a very subordinate matter. He made artificial weapons for defence and offence far superior to any that could be supplied him by nature, and by intelligent use of them he became far more efficient than any amount of muscular energy could have made him. Now nature never supplies animals with organs for which they have no use, and therefore the development of man's body practically ceased with the beginning of the development of his mind. His hand, it is true, has become more delicate, since that is an organ of great use to his intelligence. Some other changes, perhaps, also appeared, but except in a few superficial features man remained in anatomy essentially as he was at the outset, while the growth of his mental powers soon produced between him and the animal kingdom a wide chasm which we cannot bridge even in imagination.

Here, then, we readily see why it is that there has

been so much confusion and disagreement in the
attempts to place man in the scheme of classification.   Relying upon his bodily structure, he is undoubtedly to be placed with the higher primates,
and he is therefore ranked with the apes in all
schemes of natural classifications.   But he differs
from all other animals in having as his essential
character the development of a new side of his
nature which is not primarily anatomical.   When in
the pre-Tertiary times certain of the vertebrates
acquired a new character connected with the reproductive system, there soon arose a type of animals
which we call mammals.   Now classification is always based on structure, and we therefore call this
new group of animals a new class, because of their
new anatomical character and from the fact that
connected with it were other changes in anatomy
which radically modified the type of animal.   When,
however, the final race acquired his new character of
mental development, we do not regard the resulting
animal as a new class, for in this case the new de-
parture did not involve any great changes in
anatomical type.   We can therefore readily sympa-
thize with both classes of naturalists, one of which
regards man as a species of the ape family, while
the other recognizes three kingdoms—animals,
plant, and man.   Either extreme of classification is
perhaps more justified than an intermediate position.

Divergence of Character Marks the Development of Animals.

With the appearance of man there is the beginning of a new law in nature, and one of great sig-

nificance. The development of mind, and especially
of the ethical * nature of man, is producing a result
in the human race which in a measure reverses the
results of the law of natural selection, and the other
laws governing animals.

As we have seen, the history of the animal kingdom is such as can be best explained in the form of
a branching tree. In accordance with the laws of
nature, the most important of which is the law of
natural selection, the descendants of any line of
animals gradually diverge from each other like the
branches of a tree. For example the descendants of
the early type of mammals gradually assumed different characters along different lines until there
was produced the abundance of mammal orders
found in the early Tertiary (9). The exact way in
which the laws of nature work to produce such a
divergence is a matter under discussion to-day by
naturalists. Darwin tried to show that his law of
natural selection was in itself sufficient to explain
this divergence. Further study makes this more
doubtful, or at all events requires the addition of
certain other factors favoring the isolation of individuals. But whatever be the difference in our ideas
of the details, there is no question that the laws of
nature under which animals live and multiply, result
in the production of what is known as divergence of
character.

Now the essential feature of a divergence of character is isolation and separation. In some way the

---

* By the development of the ethical nature we would not mean to
imply that conscience has been evolved from the other attributes of
man, but simply that since the appearance of man the ethical element
has developed to a higher grade.

descendants of one pair of animals become associated into groups. Each group has characters of its own, and since it remains isolated from the others, at least so far as interbreeding is concerned, it soon establishes for itself a distinct line of descent, a new race or a new species. Later in the history perhaps its own descendants in like manner become separated into still other groups, and so on, the divergence becoming wider and wider as the centuries roll by. Without some sort of isolation and separation into groups divergence of character is impossible.

Convergence of Character the Result of Human Development.

The mind and ethical nature of man, and especially the law of Christ, under which man is slowly learning to live, is producing a slow but sure modification of the history of development. *Instead of producing isolation and divergence, they are producing union and convergence.*

Man is distinctly a social animal. Among the lower animals there are some that have what are known as social instincts. Instead of living as isolated individuals they associate in groups for mutual protection. Among animals of low intelligence, like fishes, this has little significance in the development. With higher animals, however, social habits produce more effect. With the higher insects, these social habits have been developed to their highest point. The marvellous series of instincts so well known among ants and bees are unquestionably

the results of these social habits. But even in insects
the social instincts are very narrow in their application, for although they produce association of individuals into colonies, although they are frequently
so far developed that the individual sacrifices his life
for the good of the colony, they do not give rise to
anything like an association of the colonies with each
other. Each colony of insects still keeps up the
natural warfare with other colonies.*

Man is also universally a social animal, and during
all his history has lived in communities. In early
history, as among savages to-day, men united for
mutual protection into communities or tribes. But
while the members of each tribe show friendship for
each other, the separate tribes have retained the
same sort of mutual enmity that is possessed by the
colonies of social animals. Hostility is the constant
relation of the tribes to each other, a hostility perhaps even more constant than among animals. This
constant hostility produced a more or less complete
separation of the tribes from each other, and the result, as in the rest of the animal world, was a gradual
separation of the tribes from each other. These are
exactly the conditions necessary for the production
of divergence in character and the formation of new
races. To this extent then the development of the
races of men was similar to the development of races
of animals.

But almost from the beginning of the development
of man we notice a new law dependent on the possession of conscience and the feeling of love. The

* To this statement there are a few exceptions.

ethical element of man's nature, though very rudi-
mentary in early history, has always belonged to
man. Its basis is love. Now love is really a new
feeling in nature, at least man is the first animal in
which it becomes of importance enough to influence
his development. Among the higher animals, such
as birds and mammals, we do find a maternal love of
the mother for her offspring, but the love ceases with
the maturity of the young. Even among social
animals there is found nothing that corresponds to
the love of one individual for another in its broader
sense. Animals combine for mutual protection but
seem perfectly indifferent to each other except in so
far as the advantage of the whole community is concerned. A soldier ant may sacrifice his life for his
colony, but he shows no feeling for another individual in distress. We would not deny that occasionally there are, even among animals, slight traces
of what must be regarded as the feeling of love in
a higher sense. It is certain, however, that such instances are few, and that the feeling of love is not
one which can be regarded as having any considerable influence upon the development of the animals.

With man, however, intelligence and conscience
have produced different conditions. Even in the
savage tribe a feeling of love for one's relatives or
friends, patriotism for one's tribe, self-sacrifice for
the good of the community, are characters which
receive the highest honor. While then with man, as
with animals, the tribal relations may be primarily
for mutual protection, it is certain that the feeling
of love for one another, the ethical element of human

nature, has had at all times the important effect of cementing the tribes into a rigid unity. Union for mutual protection is impossible without it. Among the Jews we early find the command to love one's neighbor as one's self. Their interpretation of the word neighbor was to be sure very narrow, but the presence of such a law shows that love was recognized as one of the noblest attributes of man.

The tribal relation was thus originally assumed for protection, but a mutual love cemented the tribes into units. By further application of the same feelings, the size of the tribes increased, and they were finally submerged into nations. Now it is the feeling of love which alone makes great nations possible. The increase in the size of tribes and nations has undoubtedly been brought about by conquest rather than by love. But it is the feeling of love alone which enables a large tribe to hold together. The tribe whose sympathies were confined within narrow limits would always disappear before the tribe whose broader love made possible the union of larger bodies. Those tribes in which this feeling of mutual love and sympathy was the broadest obtained the mastery over the others and increased at the expense of the others. Now since all tribes and nations have recognized a demand on man to love other members of his own nation, the scope of man's obligation to love his fellow men has constantly expanded with the increase in the size of nations. Finally, with Christ there was announced the complete law for man, the law of universal love. By Christ was man's obligation to love extended to his

enemies, by him was the word neighbor defined so
as to include every one who needed help, whether of
the same tribe or nation or another, whether friend
or foe.   Thus it was that love, the special attribute
of man, was so broadened as to include all mankind
and to bring about a universal brotherhood.   This
law of Christ looks towards the destruction of the
tribal relation, and the national relation as well, and
when it is fully established as the law of man, it will
produce one nation, one association, which shall
combine all of mankind into one union of mutual
assistance and love.   We are far enough from such a
condition at present.   It is the millenium of which
we sometimes dream, and toward which our progress
seems slow enough.   But the development of man
is tending in this direction, now that he has once
recognized that universal love is the law of his life.

Looking forward then into the coming centuries
we see the vegetable world remaining practically as
it is, except as it is modified by the interference of
man in exterminating plants not of value to him,
and improving those which he uses.   We see the
animal kingdom on the land largely exterminated by
the power of man, though life in the ocean may for
a long time remain unchanged.   But the marine
types offer no chance for the future, for they are all
low ones which reached their culmination in the past
ages.   But we see mankind left, the creature of God,
advancing in intelligence, knowledge, and morality,
to the end which we do not see and cannot imagine.
Whether there be a phase in our nature superior to
mind, which shall in the future ages be brought into

expansion and produce a new race of beings we cannot tell, but the era of *animal* life ended when that of *man* began.

It was not to be expected, of course, that this law as announced by Christ would be at first understood. The human race in his day had hardly entered into the conception of the beauty of love in its narrower sense of love to one's neighbor, and it was certainly not ready to accept the definition of neighbor as including one's enemies. It is not our purpose here to enter into a consideration of the history of the slow growth of this new law. Even yet we fail to accept it, for the successful general receives our highest honors, and the soldier is our greatest hero. We even fail to *understand* the law. The larger our nations grow, the more comprehensive become our obligations, but we do not yet believe that the time will come when the Chinaman, the African, and the Caucasian will actually unite into a common brotherhood.

All this lies beyond our present purpose. The relation of the new law to the history of life, however, is very important for us to understand. This law, as soon as it is applied to human life, produces no longer *divergence* of character but *convergence.* We have seen that the essence of divergence is separation and isolation of groups of individuals. The races of mankind which we find to-day have been produced by such a lack of intercourse among the tribes of men and by the constant enmity and warfare of early nations which have tended to keep up a constant isolation. But with the broadening of man's application of his obligation to love his neighbor, the nations become larger, and the increase in the size of the nations acts against the increase in their diversity. As fast as the members of a nation become cemented together, so fast do they begin to assume common characters. The American nation is absorbing into itself a great variety of people, perhaps faster than it can assimilate them. But as fast as they are absorbed, just so fast does the character of the nation change. The American nation no longer possesses the character of the men who struggled for independence a century ago. The history of civilization has been always marked by the absorption of the small races into the larger ones. It is only enmity and a narrow patriotic love that prevents all of the smaller nations of Europe from becoming parts of the larger ones. Thus it is that to-day the formation of new races has been checked, at least among the higher classes of mankind. The nations are growing larger, their hostilities are becoming lessened, intercourse of commerce and friendship between them is increasing, and with all this a tendency toward unification and concentration is plainly seen. *Not divergence but convergence of type is the history of to-day.*

It is therefore plain that the ethical nature of man is producing a new phase in the development of the world. It is checking the tendency to formation of new types, and is tending to unite into one the members of the race. To-day intelligence is uniting all men into closer and closer relations of commerce and education. In the future we can see it destroying all desire of conflict and victory; we can see it doing away with race prejudice, and we can see a united mankind advancing to higher and higher planes. Thus it is that the intelligence, conscience, and social habits of man, and above all his approximation toward the law of Christ, have produced a new era in the history of life, and, as a result, the development of mankind in the future is not to run parallel to the development of animals in the past. No longer are we to find a divergence and production of numerous species of animals or even numerous species of intelligence. There is to be *one* human race, a race of marvellous complexity, advancing to higher and higher planes. Mankind is to remain a unit, and so long as his chief character is the development of his intellect, he will still continue to be man, whatever be the changes that the future may see either in his physical or mental attributes.

Each Type a Master of the Preceding.

Before leaving this sketch on the outline of history, there are several other points of general interest to be mentioned. First we must notice that in all of this history the predominant animal type of any age is, as a rule, more than master for all of the animals that have preceded it. With their hard shells the mollusks have no fear of the lower animals. The activity of the articulates makes them more than a match for the mollusks or other lower animals. The vertebrates are always superior to the invertebrates, and so, as a rule, the amphibians, reptiles, birds, and mammals are seen in turn possessed of powers that

make them masters of all the lower forms. Finally,
man supersedes them all.

The Rarity of Terrestrial Life.

Again we must notice how few are the types
of animals that have ever succeeded in adapting
themselves to a life on the land. Beyond the
vertebrates and the air-breathing articulates (insects,
spiders, etc.), there are almost no land animals.
Among the mollusks there are a few snails which
have become adapted to a thoroughly terrestrial
life. There are indeed quite a number of species of
snails thus breathing air, but they are all closely
allied to each other, and commonly have a much
restricted habitat. Evidently, then, only a few
mollusks have really been able to acquire the power
of breathing air. None of the other invertebrates
have been able to acquire the power of living in the
air, and we may say, therefore, that there are only
two types of animal structure which are adapted to
life out of the water. The Protozoa all require water
as a medium through which they can feed and carry
on respiration. Cœlentera have a body too soft to
resist gravity, unless assisted by the buoyancy of the
water. The same may be said of the Echinodermata,
together with the fact that their characteristic system of organs, the water system, requires the presence
of water to make it of any value. The mollusks are
too clumsy, as a rule, to be adapted to terrestrial
life. The Vermes mostly respire through the surface
of the body or by gills, and either method makes it
necessary for them to have the exterior of their body

constantly moist. They are therefore aquatic, or occasionally live in moist earth. The Articulata and Vertebrata alone are well adapted to a terrestrial life, and of these immense groups only five classes have really become terrestrial animals (Insecta, Arachnida, Reptila, Aves, Mammalia). Five classes, then, out of the whole animal kingdom are all that have acquired the power of living in the air. These classes have more to contend with than the aquatic animals, since they have gravity to resist, and must accommodate themselves to the climate ; they must adapt themselves to varying temperature, and to many other conditions to which aquatic animals are not subjected. Now it is a law of nature that the being which surmounts the greatest obstacles is the one to rise to the highest plane. It is no wonder, then, that these five classes of animals have become predominant types, and surpass the other animals in variety and numbers. They have had the whole land in their possession.

We cannot definitely say when the terrestrial fauna appeared. In the Silurian (2) there were certainly in existence some scorpions, and probably, therefore, •other land animals. Indeed, traces of insects have been found in these rocks. Terrestrial vertebrates did not appear until the end of the Carboniferous (4), and not in any very great abundance until the Mesozoic (5-8). From the beginning of the Mesozoic, however, terrestrial fauna has played the most important part in the history of the world, and in the more recent times it has been the terrestrial forms of life to which evolution has been chiefly confined.

11

It is also a fact of great interest that both of these
types of animals which acquired terrestrial habits, the
vertebrates and articulate types, have ended their
history with the development of mind.   The development of insects has ended in the production of the
complicated insects familiar to every one acquainted
with the habits of the bees and ants.   The development of the vertebrates produces the intelligence
which has reached its culmination in man.   Both
types of terrestrial animals have developed the nervous system, and in each type the mental nature is
the special character of the highest orders.   It will
also be noticed that the development of instincts
among insects has been entirely independent of the
development of the intelligence of the vertebrates.
The two have not even progressed in parallel lines.
Each group has developed the functions of the nervous system, but one has developed the reflex functions to an extreme (instinct), and the other the
reasoning powers.   The intelligence of the vertebrates cannot then be regarded as a more highly
developed condition of the instincts of insects,
although both intelligence and instinct are functions of the mental nature.   That terrestrial life has
had the effect of stimulating both types of land
animals into the development of two distinct types
of mind, becomes therefore a matter of even greater
interest.

The Modern Fauna an Impoverished One.

Finally, we must notice that the present age is one
of short duration and comparatively impoverished

fauna. The Quaternary (10) has certainly lasted a great many thousands of years, but in comparison with the immense periods of the earlier ages, it is very short. It was with the Quaternary, indeed with the later part of the Quaternary, that the strictly modern fauna (*i. e.* the modern species) appeared, and we may say, therefore, that the duration of the present geological age is very short compared with the immense periods of time that have preceded it.

The modern fauna is regarded as an impoverished one. Of course there seem to be more animals and more species to-day than ever before, but this is due largely to the fact that we have the animals themselves to study to-day in profusion, while of the past we have only here and there a specimen. In variety of form the geological ages certainly surpassed the present. There are but few classes in which there has not been such a great extinction of orders in the past that the representatives remaining are to be regarded as fragments. Most of the invertebrates reached their culmination in the geological ages, and many of them have been for a long time on the decline. Even of the vertebrates, every class has been in much higher state of development in the past than at present. The fishes belonged to the Devonian (3), though one order has subsequently greatly expanded since then (bony fishes, in the Cretaceous (8). The amphibians belonged to the Triassic (6), the reptiles to the Jurassic (7), while the mammals existed in greater profusion in the Tertiary than they do to-day. More than one third of the orders of the vertebrates have become extinct, and of those that remain a larger number have become reduced to a few unimportant representatives of orders which in former times were abundant in number and diversity of form. Some of the orders that are still left, it is true, are perhaps more abundant, so far as number of species is concerned, than any orders of the past, but as a whole the fauna of to-day consists of remnants of the past.

Summary of the Last Two Chapters.

We may compare the history of the whole animal world to the growth of a giant tree. As members of the human race our position is among the topmost branches, and it is only by peeping down through the foliage that we dimly get an idea of the rest of the tree. The foliage is dense and our vision is obscured. Absorbed in that which immediately surrounds us we fail to see the size and magnificence of this tree of life, and, indeed, not infrequently we are inclined to believe that man stands alone, having nothing to do with the rest of the tree. Only by shutting our eyes to the dense foliage that surrounds us, and by studying the past do we learn that mankind too is a member of the same tree of life that has lived through the ages and has suffered so much from the storms of the centuries to make room for his final appearance.

If we can imagine ourselves as removed from our natural position among the branches and viewing this tree of life in perspective from the distance, it will appear something as follows. Its trunk is hidden from our view by the primeval mists of the early ages, though its existence in the dim past can be inferred from the convergence of the great branches as they approach this region of obscurity. All above the top of the trunk, however, is in sight, and we can see the tree growing with an ever widening expanse of its branches. But the storms of the ages have played great havoc. Many limbs have been torn off completely, many more have been so shattered that only a remnant is left to mark the position of a once mighty member. The prunings that have thus occurred have frequently served to give more room to the branches that are left, and these have taken advantage of the opportunity to develop large numbers of small twigs and fill the space formerly occupied by a fallen member. Occasionally we see a branch that has grown with marvellous rapidity for a short time and then suddenly died ; or another that grew for a long period as a single trunk and then suddenly expanded into minor branches and twigs. This tree of life is old and most of it has long since spent its energy. Many of the branches are dead, while others continue to live only in the shape of scraggy twigs. Death and destruction have played such havoc that the tree has become very unsymmetrical and seems from our distant view almost a wreck. But still some of the branches that are left alive are very vigorous, and when we look simply at the top of the tree the vigor displayed there almost conceals from us the wreck of the past. There at the top we see a single branch of this venerable tree that has in recent times begun to grow to an enormous size. A new law regulating its growth has stimulated it in a new direction and caused it to develop foliage at the expense of branches. So rapidly is this branch growing to-day that it bids fair to absorb into itself all of the vitality of the whole trunk, crowding out of existence its neighbors and allowing only such of the lower branches to exist as do not come in direct conflict with it. This vigorous branch is man, and although it is far above all the others, although it has expanded far more than the rest and grown so far away from the trunk that its connection with the great tree of life is sometimes obscured, still our study by the light of history shows us that this branch, too, is part of the same tree to which other animals belong, and is simply the crowning top of this tree of the ages.

CHAPTER VII.

HISTORY OF PLANTS.

THE history of the plant kingdom from early ages to the present time requires but brief notice. The problems connected with this kingdom are much simpler than those relating to animals. While plants certainly constitute important factors in the development of life, their relation to the history of mankind, which is after all the primal object of study, is very distant. Like animals they have had a history, and it is one of even more constant progression. Very early in the history of life plants became separated from animals by acquiring the power of utilizing sunlight as a source of energy, and though some of them have subsequently lost this power, it is nevertheless probable that this was the real point of separation of the two kingdoms. Whether plants or animals were the first to appear cannot be determined, though it is certain that animals, as they exist to-day, could not have preceded plants. We have seen, however, that probably neither of them preceded the other, and that the first organic life was neither animal nor plant.

Of the early history of plants we know little or nothing. The general simplicity of their structure and their lack of any complicated system of organs makes their embryological history less significant than that of animals. We can, it is true, determine that all plants start their history as single cells, and this would seem to indicate that, like animals, they have been originally derived from unicellular ancestors. But it is impossible to find traces of anything in the history of plants that correspond to the Gastræa of the animal world, nothing that can be regarded as the central starting-point from which the various groups diverged. Indeed, the history of plants seems more like the history of a single developing line of life than of a series of diverging lines.

Although embryology gives us little help in studying the history of plants, still by combining its teachings with the facts derived from comparative anatomy, some few points can be determined in regard to the line of descent through which plants have come. That the unicellular plants were followed by the low algæ cannot be questioned. That the higher plants were derived from these seems also sure. The mosses, ferns, club mosses, cycads, cone bearers (pines), and angiosperms (common flowering plants) follow each other in something like the order given, with increasing grades of structure. Now in the embryology of the cone bearers (gymnosperms) we can still find traces of an earlier stage in the history of plants corresponding to the club mosses, and even in the highest plants of all, the angiosperms, there are indications of a like history. It is significant to find that the fossil history of

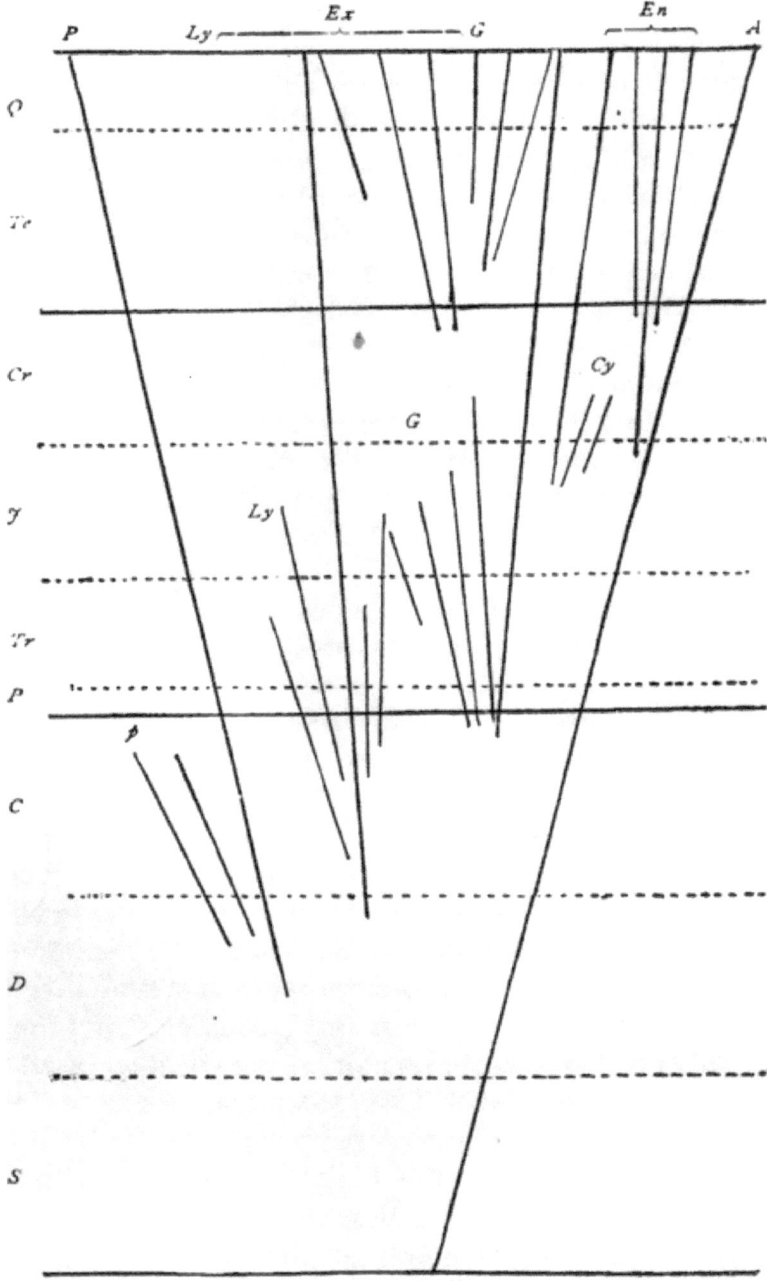

FIG. 21. Diagram illustrating the history of plants—*P* Pteridophyta. *Ly* Lycopoda. *G* Gymnosperma. *Ex* Exogens. *En* Endogens. *A* Algæ.

plants tells somewhat of a similar story, and we may thus conclude that even in plants, embryology repeats past history, at least in a measure. Of the early history of plants, however, we know practically nothing, and of the history of later ages it is necessary to combine all sources of evidence in order to get anything like a connected account.

Plants have not left so complete a fossil record as have animals. Particularly is this true of the lower forms which almost universally agree in having no hard parts adapted for preservation. So imperfect is their preservation that it is impossible in many cases to determine whether a given specimen in the earliest rocks is a plant or simply a crystallization, a worm track, or mud crack. There is no doubt that plants were in existence during the long Archean (1) age. The immense beds of graphite belonging to these times give us almost certain proof of the fact. But no traces of them except the graphite beds have survived the metamorphosis of the rocks.

In the Silurian (2) age, however, plants were undoubtedly abundant, and our first record of them is thus nearly contemporaneous with that of animals. But the plants were all of the lower types.

Marine algæ were doubtless in great numbers in the seas of this period, and everything seems to indicate that they were not unlike their descendants of to-day, which form the slimes of fresh water and some of the sea mosses. Their poor preservation makes it impossible to say much about them. During the Silurian (2) there were also in existence at least two types of land plants, living perhaps in swamps or shallow water. There was a delicate little plant named Psilophyton, thought to be a link between two later groups (rhizocarps and lycopods), and also a plant of a size so large as to resemble a tree, though in structure of so low a type as to belong to the algæ (Nematophyton). Other land plants were living also, but they were all of a very low type. The land vegetation was marvellously different from that of to-day, and we may almost look upon it as a flora of marine algæ which had been transferred to the land and then enlarged. Still there was, even at this time, a differentiation into stem and leaf, and this does not occur in ordinary algæ. The Silurian flora was in one marked respect very different from the Silurian fauna. The latter has surprised us with its diversity and high grade of development, while the former may equally surprise us with its scarcity and its low grade. The development of plants had not reached such a high state as the development of animals at this time.

In the next age (Devonian) there was an undoubted advance. Algæ were still in abundance, as indeed they have been in all ages up to the present time. But with the Devonian, undoubtedly, higher plants appeared. True rhizocarps and lycopods (horse tails) were abundant at this time, though possibly they may have begun in the previous age. In the Devonian also we find true ferns, some of them small and others of large size like our tree ferns. In this age, too, appeared the first indication of the flowering plants in the form of gymnosperms or conifers (yews and cordiates, an extinct family). The flora of the Devonian was thus on the whole composed of the highest of the cryptogams (lycopods) and the lowest of the phanerogams (evergreens and conifers).

In the interval between the Devonian (3) and the Carboniferous (4) ages there were great changes in the level of the land. After various submergences it finally arose clothed with a flora of precisely the same general character as that of the Devonian, but with new species and genera. The changes had been sufficient to obliterate old and produce new ones. The flora still consisted of the same groups of plants with no advance in structure. The ferns became very diversified, and the horse tails and lycopods reached their greatest size, but there was no advance in structure. The age was characterized, however, by the great development of the cone bearers (Gymnosperma), ferns (Pteridophyta), and the horse-tail group (Lycopoda). The Carboniferous was an age of especially abundant vegetation, and to the plant growth of that time we owe our beds of coal.

Coming now into later periods, we find with the beginning of the Mesozoic again a new flora, but again no advance in structure of much importance. The typical vegetation of the Carboniferous became less prominent and the cycads appeared. The cycads are the prominent type of the Mesozoic. During the passage of the Mesozoic, however, we see a change in the flora of a radical sort. Even with the Triassic (6) the giant horse tails began to wane, and the more modern forms of moderate size appeared. In the Jurassic (7) the cone-bearing plants reached their highest development, from which culminating point they have steadily declined. Some of our modern forms arose at this time, the genus Pinus being in considerable abundance. During the Jurassic also the first of the endogens made their appearance in the form of screw pines and grasses, and a little later the palms appeared.

The Cretaceous (8) period is, however, the line that marks the boundary between the older vegetable world and the modern, just as the beginning of the Mesozoic separated the older animals from the modern forms. Occasional traces of the higher flowering plants are found in the lower rocks of the Cretaceous, but it is in those of the upper Cretaceous that they appear in abundance. Here there suddenly bursts upon our view an abundant flora of modern forms. Most of the plants of that time belonged to the same genera as those living to-day, and many of them were, so far as we can tell, of the same species. Our knowledge of them is chiefly confined to the preserved leaves, and these are a rather unsatisfactory basis for determination of species, but there seems to be no doubt that the poplars, willows, beeches, oaks, birches, alders, laurels, sassafras, magnolias, butternuts, hickorys, and many others were represented in profusion. There is also pretty good evidence of the existence of the higher flowering plants, like the Compositæ, at this time, though doubtless the plants with inconspicuous flowers predominated. In short, at this time the modern botanical world was nearly complete, so far as type was concerned. Its highest forms had been developed even then, and from that time to the present the growth has been simply in the expansion of types then existing, and in the relative increase of the numbers of the highest flowering plants. The gymnosperms certainly have become less abundant, and the plants with large conspicuous flowers have greatly increased in preponderance.

The endogens have remained nearly constant since then, so that we may say that in the later Cretaceous the vegetable world reached nearly as high a position as it has to-day.

We may now ask whether in this history of plants we can trace the origin of the different groups of plants from each other. As already pointed out, the similarity in structure would lead us to believe that the Silurian land plants were simply terrestrial forms of the marine algæ. The same conclusion would follow from the study of their anatomy and development. The rhizocarps, to which some of the early land plants seem to be related, are themselves closely related to various algæ. So, too, may the ferns and horse tails or the Carboniferous be regarded as derived from lower marine plants (algæ), with various modifications, probably through some lost intermediate steps. The cone bearers, however, give evidence of having descended not from the algæ, but from some high form of plant belonging to the Lycopoda. Their structure, and especially their method of reproduction and development, shows them closely related to certain plants classed with the Lycopoda (Isoetes), but differing from the ordinary Lycopoda in having two kinds of spores, a large one and a small one. These micro- and macrospores correspond to the pollen and embryo sac of the flowering plants, and from some such source doubtless have the latter been produced. The endogens again were unquestionably not products of the cone bearers, but they had an independent origin down the main line, not unlikely an origin close to that of the gymnosperms. The higher flowering plants, the exogens, though closely related to the gymnosperms, were probably not derived from them directly, and were certainly not derived from the endogens. Probably they had an independent origin from some lower group close to the gymnosperms. It would seem then that some low algæ-like type of plant at one time became terrestrial, and then there occurred a divergence from it which resulted in the various forms of cryptogams, the mosses, ferns, and lycopods. These groups soon expanded and reached their culmination in the Carboniferous. But along one line of descent (Isoetes, etc.,) two kinds of spores were produced, and this line now expanded and produced the flowering plants, the gymnosperms, endogens, exogens, all of which plants retain the two kinds of spores (pollen and embryo sac), but seem to give indication of having had independent origins from the cryptogams.

It is a fact of no little interest that the flora of the world probably did not, as would be expected, originate in the tropical regions, but on the contrary in the Arctic zone. Abundant evidence shows that in the northern regions the various groups of plants first appeared and culminated. This must, of course, indicate that the climate of the Arctic zone was not the frigid one that it is to-day.

The first point that strikes out attention in this history, is that the life of plants has been more evidently one of progression than that of animals. We do not find that the Silurian (2) age opened with an abundant and highly specialized flora to correspond to the highly developed fauna of the times. The plants were few and all of the lower orders. Comparatively speaking, then, we may say that the animal kingdom had reached a much higher state of development by the beginning of the Silurian than the vegetable kingdom. While all of the classes of animals (except two) had appeared at this time or during the Silurian age, there were no plants higher than rhizocarps, and nearly all of the plants in existence were algæ or still lower types. The whole vegetable world which is familiar to us to-day was still to be developed.

We notice next that although in the early periods the animal world developed faster than the vegetable world, still we find that before the close of the geological ages the relation was reversed. The vegetable kingdom reached its culmination long before the animal kingdom. With the Cretaceous (8), the vegetable world had developed its highest types, the subsequent history being only an elaboration of them. But at that time the highest class of animals had not appeared at all, for the true mammals came into existence only in the next age (Tertiary), and very great advance took place in them even later.

In general it remains to be noticed that with the opening of the Silurian, the vegetable kingdom had reached a condition where cellular differentiation was found. All the powers pertaining to the cell seemed to be fixed, for there is no reason for thinking that the cells of early plants were not about like those of to-day. During the Silurian also the stem and leaf were differentiated. Sexual reproduction probably had begun at this time, but it was in a low form, and the formation of fruit did not appear until later. Thus the advance has been almost wholly in refinements of parts, and not in the productions of new features.

Finally, we notice that it is impossible at present to trace the various types in the vegetable kingdom to a common centre. We do not find any evidence of a rapid divergence from one central point of origin. In the animals we have reason for thinking that all of the great types arose early as branchings from one simple type, the Gastræa, and that the different sub-kingdoms did not arise from each other to any noticeable extent. In the vegetable kingdom the reverse seems to be nearer the truth. New types of plants were constantly arising till the Cretaceous, and there is every reason for thinking that they arose from the earliest existing plants in all cases. In other words, the history of the vegetable world is rather to be compared to the history of a single sub-kingdom of animals than to the history of the whole animal kingdom. Like the vertebrates they have had most of their development since the Silurian age, although it is true that their history really began earlier. In the fossil history of plants, therefore, we find just such an advance as we have seen in the vertebrates, and not such a mixture of ad-

vance and stationary condition as we have seen in
the animal world taken as a whole.   This difference
is certainly a striking one, as we take a cursory
glance at the fossil study of plants and animals as
they are known to-day.   The fact may, however, be
partly due to our incomplete knowledge of the lower
orders of plants.   The lowest classes of plants are
almost wholly unrepresented by fossils, and through
all the geological ages it is the higher classes which
have been most preserved and most studied.   Perhaps if our fossil record of lower plants were as complete as that of the lower animals, we should find
that here too there has been divergence from common centres to a much greater extent than now
appears.   But however that may be, the facts as
collected at the present time point to a history of
much more continuous progression among plants
than animals. See Fig. 21.

CHAPTER VIII.

THE FUTURE OF THE LIVING WORLD.*

SCIENCE is at all times trying to read the future by means of the past. No one questions the right of astronomy to make predictions, and her success in this direction is everywhere recognized. It is, then, certainly a legitimate question for us to ask here whether the past history of life cannot give us indications of the direction of the drift of the living world, and thus enable us with something like probability to look into its future. Such predictions cannot of course claim anything like the certainty of the predictions of astronomy, for the complexity of the problem is too great. At the same time, a little study will show that there are some definite results plainly indicated by the drift of the past, and a clearer idea of the meaning of past history can be obtained by trying to see in what direction it turns our thoughts for the future.

We have just seen that a fair idea of the life of the world can be obtained by comparing it to a giant

* The ideas advanced in this chapter were first published in the American Naturalist, 1886, from which part of the following pages are quoted.

tree.   In this comparison we have seen that the tree
is to be regarded as an old one, all of whose branches
show by their shattered condition, the effects of the
storms of the ages.   Comparatively few branches
remain alive, while a larger number have either disappeared or become reduced to a few still vigorous
shoots.   The highest branch alone appears to be in
its primal vigor, still rapidly growing and expanding,
and this because of the influence of a new life principle, perhaps engrafted into the old tree.

Now if such a comparison is a correct one, it
is evident that the tree must be looked upon as
being near its death.   Of course, however, it is possible to question the correctness of this comparison,
and we must therefore ask whether there are really
any grounds for believing that the life of the world
has passed its prime, that while man is the crowning
creation, his appearance indicates the decline of the
living world as a whole.

Development More Rapid in Early Ages.

In order to answer this question in the affirmative,
it will only be necessary to refer to some of the facts
already noticed regarding the previous ages.   First
we may notice again the significant fact that the
development of the animal kingdom seems to have
been more rapid in the earliest times than it has
been in subsequent ages.   The diversity of the Silurian (2) fauna has already been noticed, and this, of
course, means that a large part of the evolution of
type had occurred previous to the Silurian age.   All

subsequent development has been only elaboration and not the production of new types.

Now we are not at liberty to assume an indefinite amount of time prior to the Silurian. Of course it is impossible to say just how long a time elapsed between the origin of life and the beginning of the Silurian, but it seems hardly possible that it could have equalled the time that has elapsed since then. But, upon evolutionary theories, the animal kingdom must have developed during that period from the lowest unicellular condition to the complex and diversified fauna of the Silurian. When we consider, therefore, that during this time all of the important groups of the animal kingdom arose (with perhaps the exception of the vertebrates) and that none have arisen since that time, it becomes quite evident that evolution must have progressed with greater rapidity at that time than it has since. This conclusion is no new one, for many naturalists have seen the necessity of making some such assumption. It will, indeed, be generally acknowledged that evolution in the earliest ages was more rapid than at present.

Here, then, we see another point of likeness in the comparison of the living world with the life of the individual. An individual when it begins life grows most rapidly, but from the very moment of its birth the rapidity of growth lessens until a stationary condition is reached at maturity. So it seems in the longer history of world life. Its growth was most rapid at the birth of life and has been decreasing since that time, not with regularity, perhaps, being frequently interrupted by periods of more rapid expansion, yet nevertheless on the whole decreasing. If this is so, it plainly implies an end to the process.

The Organic World Approaching a Limit.

That the organic world is approaching a limit in its development is a conclusion which does not, however, depend upon any vague idea of growth for its support; for a little thought upon discovered laws will clearly show us that there must be a limit to advance. The best definition which has ever been given to the grade of structure of animals and plants is the degree to which differentiation of organs is carried. Evolution as it tends to raise the grade of animals is constantly increasing the amount of differentiation and specialization. We have already seen that such differentiation and specialization becomes self-limited. A low undifferentiated and unspecialized organism has an indefinite possibility in its lines of specialization. A simple spherical cup of cells, the supposed common ancestor of the animal kingdom, may be modified in a very great variety of directions, each of which may give rise to a different type of animal. This possibility lies in the fact that it is as yet undifferentiated and unspecialized. But just as soon as it does become modified in any one direction the possibilities decrease. Some of the descendants of this ancestor becoming vertebrates are forever precluded from becoming anything else; others becoming mollusks must remain mollusks forever, with all of their descendants. And as later descendants become further modified in any direction into definite types the chance for future modification becomes rapidly less. It is only the absolutely undifferentiated which has infinite possibilities, for as soon as a single step is taken in any direction, the possibilities become finite. Now it is plain that this continued specialization cannot go on forever. Since evolution does not retrace its steps, every step in advance limits the possible lines of development. All the descendants of the vertebrate line must conform to the vertebrate type. The vertebrates become separated into fish, reptile, and mammal, and each group is still further fettered in its development by the special line which its ancestors have taken. The descendants of the animals which have started the order of birds cannot take any new line. They can develop this type to perfection, or they may lose their special characters, but there they must stop. And thus, with every step in advance, the possibilities of expansion are constantly decreasing.

Now a continued specialization of this sort is sure to reach a limit ; it must run to extremes and eventually stop. Physical laws will of themselves set limits to every line of advance, even if there be no such limits determined by the organism itself. It is easy to find examples which will show that such has been the general history of groups in the past. Some have reached the extreme of their development in the distant past, and have ceased to advance, or, perhaps, have disappeared. Others seem even now to be at the summit of their advance, and others still are yet advancing. The line of development represented by the trilobites has completely ex-

hausted itself.   It rapidly approached its limits even
in the Silurian (2), and then began to dwindle away,
and has disappeared entirely.   The brachiopods had
also at this time reached their point of highest
specialization, and had become a highly developed
group even at this early age.   Since then they have
remained stationary as to their organization, having
steadily decreased in numbers, and the few that are
left show no advance over the Silurian forms.   The
cephalopod mollusks gradually increased in complexity during the Paleozoic (2–4) ; and finally a
limit of the shelled forms was reached in the ammonites of the Jurassic (7) and Cretaceous (8).   The
culmination was followed by extinction.   Meantime
a second line of development began, that of the
naked cephalopods (squids, cuttle fishes), and this
has gone on advancing until the present time.   The
decapod Crustacea represent a group which is even
now near its culmination.   From their first appearance in the Carboniferous (4) there has been a tendency to a concentration of organs toward the head.
As this specialization advanced, the abdomen became smaller, while the head region became larger.
Finally, in the crabs, everything was concentrated
in the head region.   The abdomen remained as little
more than rudiment.   Evidently we are here near a
limit, and we may look upon the crabs of to-day as
the culmination of the special line of development
which has characterized this line of animals.   The
vertebrates in general have been continually advancing during geological times, with a continued increase
in specialization and in multitude of types.   But even

here there has been the same story of limitation.
The ganoids culminated in the Devonian (3), and
have advanced no farther. One great line of reptiles
reached its limit in the Jurassic (7). And so everywhere. The study of every group teaches that the
past history has been a gradual specialization which
approaches a limit. In many cases in the past this
limit has been reached, and advance has ceased ; in
others, the animals are still on the road toward it.

It is plain, then, that unless there is some way for
new lines of specialization to arise, the evolution of
the animal world is inevitably approaching its end.
With every advance in differentiation, the possible
lines of development decrease, and since the actual
lines followed are tending to run themselves out, the
whole must eventually stop. Is it, however, possible
that new lines of differentiation may arise, and thus
the development of the living world go on indefinitely ?

New Lines of Specialization Not Now Appearing.

First, we must notice that the development of the
vegetable kingdom practically ceased long ago. As
already noticed, the Cretaceous (8) rocks show us
representatives of the highest orders of plants. At
that time the plants in existence do not seem to have
differed materially from those of to-day, since many
species are identical so far as can be determined.
The vegetable kingdom thus practically reached its
culmination at this time ; for although many new
species have appeared since then, there has been no
advance in type. The time since then has been long

enough certainly for a great amount of change, if
the limit had not been practically reached. It has
been sufficient for the entire development of the
class of mammals, with its great profusion ; but the
plant world has remained nearly stationary, and this
suggests to us that we may look for no farther development along this line.

With the animal kingdom, however, it is different.
Even until the present era we can see that development of new and higher forms has continued. Is
there any indication that it has reached its end ?

That there is a theoretical possibility of the origin
of new types cannot be denied. New types, *i. e., new
lines for specialization*, can arise only from undifferentiated forms. But such undifferentiated forms
still exist in great numbers. Even the most unspecialized forms of all, the unicellular animals, are
abundant enough, and in all groups we are acquainted
with more or less generalized types. Theoretically,
then, there is no reason why any one of these forms
should not expand itself, and thus form an eternal
source of new world forms. So long as the unspecialized forms do not become extinct, we cannot
deny the possibility of an infinite number of future
sub-kingdoms, which would of course make the
animal kingdom an example of never-ending evolution. But all of our evidence indicates that such
a future is probably not a practical possibility,
even though, so far as we can see, it may be
a theoretical one. All biological studies point
strongly to the conclusion that, instead of several
points of origin the animal kingdom has had only

one. The sub-kingdoms have not arisen independently from the Protozoa, but have all had a common ancestor, the Gastræa, and this means that only once has the unicellular form given rise to an important line of multicellular descendants which perpetuated itself. Though the Cœlentera stand very near this primitive Gastræa, there is no evidence that the sub-kingdom has the power of further production of new types ; but, on the contrary, everything tends to show that, whatsoever differentiation of this simple type ever did take place to give rise to the sub-kingdoms, occurred before the Silurian. Since paleontology shows that no great types have arisen since the Silurian, it is plain that all of the expansion of the simple unicellular form must have taken place before the Silurian. And coming through the later ages, we find that the evidence is the same in its tenor. The conclusion everywhere seems to be that when a generalized form has given rise to one or two lines of development, it either disappears or loses its power to originate new forms. Every bit of evidence which indicates a fundamental unity of the animal kingdom testifies to the same. Without questioning the theoretical possibility that any or all of the existing unspecialized forms may in the future develop, we must acknowledge that the probability is against it. Nothing in history indicates that these groups retain the power to expand, and there is, therefore, no reason for thinking it a possibility in the future. Remembering what a large number of groups we are learning to trace back to the Silurian, remembering that development

in the later geological ages has consisted simply
in the expansion of groups appearing long before,
we must conclude that the power of the undifferentiated forms to expand into different lines of development disappears very early in their history.
While then we cannot deny the possibility of an
indefinite future development from the existing
generalized types, it is certainly improbable that
any new great groups will arise.   Man, seizing upon
the last undifferentiated faculty, the intellect, is
developing this to the extreme. With him, the
animal world proper has ended and the intellectual
world begun.

If there is anything further needed to convince us
that the evolution of animals ceases with man, we
have only to notice his influence upon the rest
of the animal kingdom.   We cannot yet compute
that influence, but it will doubtless be the death-
blow to the evolution of animals.   Man is rapidly
causing the extinction of almost all land animals, at
least the larger ones.   As the frontiers of civilization
are being extended farther and farther into the
uninhabited regions, he is driving out of existence
all of the large animals and many of the smaller
ones.   We have only to look ahead a comparatively
short time to see the extinction of nearly all land
animals, except such as may strike man's fancy to
use or preserve.   To what extent this may apply to
other animals—to insects, marine animals, etc.—is
not clear.   But in the highest group of animals, the
vertebrates, it is pretty clear that man is eventually
to bring about not only the end of advance, but also

the practical extermination of all animals except such as he especially preserves.

His mastery over the higher vertebrates is so unbounded that he is the only one who has any possibility of farther advance. With the lower animals, his competition is not so severe, but, as we have seen, these lower types have practically ended their evolution in the distant past.

To man alone, then, are open further possibilities of higher development, and his development must be wholly mental. His unique position in nature comes upon us now with double significance. Even more forcibly than ever do we see the significance of the new law of love to which he is trying to adapt his life. Remembering how lines of specialization exhaust themselves, and how divergence of character ends in extremes, we begin to get a grand conception of the law of love which binds mankind into unity, thus checking the development of castes and forcing him to advance as a whole. The law of love is the only law which could produce the highest mental and moral development.

REFERENCES.

A long list of references would be inappropriate in this work. The following short list may be of value to those wishing further reading upon the topics treated in the foregoing pages :

PROTOPLASM AND CELLS.

BERTHOLD. "Studien über Protoplasma mechanik," Leipzig, 1887.

NELSON, JULIUS. "Heredity and Sex," *American Naturalist*, 1887.

PASTEUR, L. "Studies on Fermentation," Macmillan & Co., 1879.

SCHWARZ. "Morphological and Chemical Composition of Proto-plasm," *Zeitr. z. Biol. d. Pflanzen*, v. 1887.

WEISSMAN. "Essays upon Heredity and Kindred Biological Problems," Clarendon Press, 1889.

WHITMAN, C. O. "Seat of Formative and Regenerative Energy," *American Jour. of Morphology*, 1888.

LIFE.

BASTIAN, CH. "The Beginnings of Life," D. Appleton & Co., 1872. "The Modes of Origin of Lowest Organisms," Macmillan & Co., 1871.

BEALE, SIR LIONEL. "The Mystery of Life," J. & A. Churchill, 1871. "Protoplasm ; or, Matter and Life," Lindsay & Blakiston, 1874.

DRYSDALE. "Protoplasmic Theory of Life," Ballière, Tyndall & Co.

HUXLEY, TH. "The Physical Basis of Life," Humboldt Library ; originally delivered as a lecture in 1868. "Anatomy of Invertebrates," D. Appleton & Co., 1878. Article on Biology in Encyclo. Britannica. "Animal Automatism," *Fortnightly Review*, 1874.

MINOT, C. S. "On the Conditions to be Filled by a Theory of Life," Proc. of Am. Association, 1879.

MORRIS, C. H. "Organic Physics," *American Naturalist*, 1882.

DuBois REYMOND. "Seven World Problems," *Pop. Sci. Monthly*, 1882.

TYNDALL, J. "Floating Matter in the Air," D. Appleton & Co., 1882.

WARD, LESTER. "Organic Compounds in their Relations to Life," *American Naturalist*, 1882.

EARLY HISTORY OF ANIMALS.

BALFOUR, F. M. "A Treatise on Comparative Embryology," Macmillan & Co., 1880.

BUTSCHLI. "Bemerkungen zur Gastraea Theorie," *Morph. Jahr.*, 1884.

CONN, H. W. "Marine Larvæ and their Relation to Adults," Biol. Studies of J. H. U. iii., 1883. "A Suggestion from Modern Embryology," *Science*, 1885.

HAECKEL, E. "Die Gastraea Theorie," *Jena Zeitschrift*, 1874-5.

HYATT, A. "Genesis of the Arietidae," Smithsonian Contributions to Knowledge, 1889. "Larval Theory of the Origin of Cellular Tissue," Proc. of Bost. Soc. Nat. Hist., 1884.

LANKESTER, E. R. "Notes on Embryology and Classification of the Animal Kingdom," *Quar. Jour. Mic. Sci.*, 1877.

McMURRICH. "The Gastræa Theory and its Successors," Biological Lectures, Ginn & Co., 1890.

MITSCHNIKOF. "Embryologische Studien an Medusen," Wien, 1886.

SEDGWICK, A. "Origin of Metameric Segmentation," *Quar. Jour. Mic. Soc.*, 1884.

WILSON, E. B. "Some Problems of Annelid Morphology," Biological Lectures, 1890, Ginn & Co.

GEOLOGICAL HISTORY.

COPE, E. D. "Origin of the Fittest," D. Appleton & Co., 1887.

DANA, R. "New Text-book of Geology," Ivison, Blakeman, Taylor, & Co., 1883.

GEIKIE, A. " Text-Book of Geology," Macmillan & Co., 1885.
LeCONTE. " Elements of Geology," D. Appleton & Co., 1882.
LYELL. " Principles of Geology," D. Appleton & Co., 1872.
NICHOLSON and LYDECKER. " A Manual of Paleontology," Wm.
    Blackwood & Sons, 1889.

PLANTS.

DAWSON. " The Geological History of Plants," D. Appleton & Co.
NICHOLSON and LYDECKER. " Manual of Paleontology," vol. ii.
SACHS. " Text-book of Botany," Clarendon Press, 1882.

MISCELLANEOUS.

CLAUS and SEDGWICK. " Elementary Text-book of Zoölogy," Mac-
    millan & Co., 1884.
CONN, H. W. " Natural Selection and Christianity," *Methodist
    Review*, 1891.
DARWIN, C. " Origin of Species," " Descent of Man," D. Apple-
    ton & Co.
FISKE, J. " The Destiny of Man," Houghton, Mifflin, & Co., 1885.
    " Outlines of Cosmic Philosophy," J. R. Osgood & Co., 1875.
HUXLEY, TH. " Evidence as to Man's Place in Nature," New York,
    1863.
LUBBOCK, J. " Pre-Historic Times," London, 1869.
MIVART, ST. GEO. " Genesis of Species," Macmillan & Co., 1871.
    " Lessons from Nature," D. Appleton & Co., 1876.
ROMANES, G. " Mental Evolution in Animals," 1884 ; " Mental
    Evolution in Man," 1889, D. Appleton & Co.
REED. " Evolution versus Involution," J. Pott & Co., 1885.
SPENCER, H. " Principles of Biology," D. Appleton & Co.
TYLOR. " Primitive Culture," John Murray, London, 1871.
WALLACE, A. " On Natural Selection," Macmillan & Co., 1870.
    " Darwinism," Macmillan & Co., 1889.

INDEX.